SpringerBriefs in Applied Sciences and Technology

For further volumes:
http://www.springer.com/series/8884

Ajey Lele

Mission Mars

India's Quest for the Red Planet

 Springer

Ajey Lele
Institute for Defence Studies and Analyses
New Delhi
India

ISSN 2191-530X   ISSN 2191-5318   (electronic)
ISBN 978-81-322-1520-2   ISBN 978-81-322-1521-9   (eBook)
DOI 10.1007/978-81-322-1521-9
Springer New Delhi Heidelberg New York Dordrecht London

Library of Congress Control Number: 2013941495

Printed on acid-free paper

Springer is part of Springer Science+Business Media (www.springer.com)

To my sister Suvarna

Preface

India's first mission to Mars is scheduled for launch in November 2013. This mission has attracted a significant amount of attention globally. At the backdrop of this mission this book attempts to discuss the interests and importance of Mars.

Reaching Mars is a quantifiable science but convincing people about the importance of Mars is more about art. This book oscillates between science and art to understand the scientific and technological endeavour called Mission Mars and its relevance to society. It is hoped that this work would offer a reader with an appreciation of contemporary happenings with regard to the human efforts to study Mars and the nature of complexities and the context in which these attempts are evolving.

In respect of India's Mars programme, the author is grateful to Dr. K. Radha-krishnan, Chairman of the Indian Space Research Organization (ISRO) for useful discussions and for giving access to requisite information with regard to India's first Mars mission. Also, I would like to specially thank Dr. V. Adimurthy, Mr. Shantanu Bhatawdekar, Mr. V. Gopalakrishnan and other ISRO scientists and Professor J. N. Goswami, Director, Physical Research Laboratory, for providing relevant information and useful discussions.

I owe my sincere gratitude to the Institute for Defence Studies and Analyses (IDSA) and Director General Dr. Arvind Gupta, for encouraging me to undertake this work. I am also grateful to Mr. Parveen Bhardwaj and Mr. Vivek Kaushik from IDSA for their support.

Lastly, I extend my gratitude to my parents, wife Pramada and son Nipun for their support. The contents of this book reflect my own personal views.

INSTITUTE FOR DEFENCE
STUDIES & ANALYSES

Ajey Lele

Contents

Acronyms

BWG	Beam Wave Guide
CA	Core-alone
CBM	Confidence Building Mechanisms
CTH	Crew Transfer Habitation
DoS	Department of Space
EIS	Electron and Ion Spectrometer
ESA	Electron Spectrum Analyser
ExoMars	Exobiology on Mars
FOV	Field of View
GSLV	Geosynchronous Satellite Launch Vehicle
IIST	Indian Institute of Space Science and Technology
IMI	Ion Mass Imager
IMP	International Mars Project
INCOSPAR	Indian National Committee for Space Research
IRNSS	Regional Navigational Satellite System
ISA	Ion Spectrum Analyser
ISRO	Indian Space Research Organisation
ISSDC	Indian Space Science Data Centre
JAXA	Japan Aerospace Exploration Agency
LAP	Lyman Alpha Photometer
LEM	Liquid Engine Motor
LEO	Low Earth Orbit
LEOS	Laboratory for Electro Optics Systems
LFA	Low Frequency Plasma Wave Analyser
LLRI	Lunar Laser Ranging Instrument
MAVEN	Mars Atmospheric and Volatile Evolution
MCC	Mars Colour Camera
MDC	Mars Dust Counter
MDRS	Mars Desert Research Station
MELOS	Mars Exploration with a Lander and Obiter
MENCA	Mars Exospheric Neutral Composition Analyser
MGF	Magnetic Field Measurement
MIC	Mars Imaging Camera

MOI	Martian Orbit Insertion
MSM	Methane Sensor for Mars
MTT	Martian Transfer Trajectory
NASA	National Aeronautics and Space Administration
NGIMS	Neutral Gas and Ion Mass Spectrometer
NMS	Neutral Mass Spectrometer
PET	Probe for Electron Temperature
PIs	Principal Investigators
POCs	Payload Operations Centres
PRL	Physical Research Laboratory
PSLV	Polar Orbiting Satellite Launch Vehicle
PWS	Plasma Wave and Sounder
RosCosmos	Russian Federal Space Agency
SCC	Spacecraft Control Centre
SOI	Sphere of Influence
SRE	Space Capsule Recovery Experiment
TGO	Trace Gas Orbiter
TIS	Thermal Imaging Spectrometer
TPA	Thermal Plasma Analyser
UAVs	Unmanned Aerial Vehicles
USGS	United States Geological Survey
UVS	Ultraviolet Imaging Spectrometer
VEGA	Venus–Earth Gravity Assist
XUV	Extra Ultraviolet Scanner

Figures

Tables

Part I
Prelude

Part 1
Prelude

Chapter 1
Introduction

> *We went to Mars not because of our technology, but because of our imagination.*
>
> Norman Cousins

Outer space (is also simply called as "space") has always caught humans' imagination. This region is a void between different celestial bodies like moon, planets, stars and asteroids to which humans are attracted for centuries. Outer space has unique (and at times even elitist) psychological significance. To achieve success with various space endeavours is far difficult and costly than to reach to any part of the earth [1]. Various nation-states regard the glory of their "celestial" achievements as a signature of their supremacy. It demonstrates their capability to take on difficult technological challenges. During the Cold War era, achievements in space have even been instrumental (in limited sense) to alter the nature of (Cold) War. In present context in order to appreciate the interests of states towards space activities, it could be useful to raise few questions. The basic query could be: "in the post-Cold War era has the policies of states changed in the space domain?" Or, "do the states still continue to recognise their achievements in space domain as symbol of their supremacy even in the twenty-first century?" In general, has the process of power rebalancing changed the ideas in regard to investing in outer space activities? Has the process of globalisation changed the "context" of investments in space? Are economic compulsions forcing states to reduce their investments in space and have obliterated the appetite for taking chances for investing into unknown? Since the scientific community was not able to replicate an act similar to the man mission to moon, have the people in general lost their interest in outer space activities?

There may not be necessarily any direct answers to such issues. This book is an effort to look for answers for some of such questions in a particular case. This work aims at understanding of the probable reasons behind modern day investments in undertaking missions to Mars at the backdrop of India's proposed Mars mission which is expected start its journey towards Mars during November 2013.

---

*American author and political journalist had once famously said this after successful Viking Mission. Viking was a pair of American space probes that provide significant amount of information during 1990 to early 2000, http://www.astrodigital.org/mars/whymars.html, accessed on Nov 1, 2012.

A. Lele, *Mission Mars*, SpringerBriefs in Applied Sciences and Technology,
DOI: 10.1007/978-81-322-1521-9_1, © The Author(s) 2014

Here, the attempt has been to realise both the operational and conceptual validity of such investments.

In order to start the enquiry of such query, it could be of interest first to appreciate the relevance of space technologies by comparing and contrasting Cold War era and the post-Cold war interests of various states in the space domain. There are few basic questions which are required to be posed to get some clarity on the modern "context" of space. First, are states investing in the post-Cold War era in space technologies for same reasons than that of in the Cold War era? Is the context strategic or socio-economical or both? Secondly, the success of the United States mission to the Moon in the Cold War era in some sense was viewed as a "victory" over the erstwhile USSR. Do such similar "notions" still persisting? Or, has the world become more realist? Alternatively, with increase in the number of space players, has the race or the tendency of one-upmanship also broadened? Third, is reaching Moon is still being viewed as a symbol of space supremacy or has Moon been replaced by Mars or some other form of activity in space? Fourthly, are there concerns about any possibility of asteroids or meteorites hitting the earth in near future and are humans keen to find scientific solutions to address such issues? Lastly, is there increasing interest to know more about the solar system in which we live in and issues like the possibility of existence of life outside the earth?

Since the beginning of space era with every passing decade, the global dependence on space assets has significantly increased, and in some sense, it could even be argued that today the earth is being ruled from the space. There has been a significant amount of interest in space sciences for various states for many years, and continuously scientists are found making efforts to find the answers to various unknowns [2]. During 1960s, the USA and erstwhile USSR had began an impressive effort to know more about our solar system. Particularly, the success of Apollo greatly helped the people and nation-states to hold a widespread interest into space sciences. Interestingly, the planetary exploration also could be viewed to have began as a race between the USA and erstwhile USSR, the way lunar exploration had. It could be said to have started with an attempt to "first" put a craft in the close vicinity of Venus [3]. However, it has been observed that within our solar system, it is the Mars which has attracted maximum attention both by scientific community and by the common man. This could be because Mars being viewed as a planet having some similarities with the earth and probably hence capable of hosting life on its surface. Naturally, with abounded of interest in knowing anything about the existing extraterrestrial life (if any) such interest continues to persist. Science fiction also could be said to have played its own role in popularising such notion.

It is important to appreciate that almost everything from satellites and space travel to nanotechnology to synthetic biology and robots existed in science fiction before they were realised in the laboratory. There have been extremely popular television serials like the Star Wars which in a way had impacted the imagination of both the common man and the scientific community. Science fiction could have been somehow beneficial for the scientists particularly in the space arena to think different and to think big.

At the same time, there have been few drawbacks with the science fiction too. The second man to walk on the moon Buzz Aldrin puts these limitations succinctly. He argued that "I blame the fantastic and unbelievable shows about space flight and rocket ships that are on today ... If you start dealing with fantasy and beaming people up and down and travelling seven times the speed of light, you are doing damage. You're not helping. You have young people who have got expectations that are far unrealistic, and you can't possibly live up to the expectations you have created in young people. Why do they get bored with the space program? That's why" [4]. Particularly, in regard to Mars mission, it could be said that expectations were raised more prematurely about the possibilities of the human habitats over the Mars. May be in a limited scene science fiction was responsible for this. It could even be argued that humans somehow appears to be obsessed with the idea of finding life on Mars and have latched on to any evidence—however improbable—that might support its existence. Charting the history of the often deceptive scientific (not to mention literary) findings made about Mars, this viewpoint suggests that our desire to find extraterrestrial life says more about the human need for companionship and communication than about the true past of the planet [5]! Any debate on Mars needs to factor these facts to appreciate that why we think of Mars mostly in a particular manner even today.

Over the years, few states have mastered the art of putting satellites into the low, medium and geostationary orbits. Such satellites which are put in space for the purposes of remote sensing, navigation, communication, etc., are positioned maximum to the height of 36,000 km above the Earth's surface. However, reaching Mars is far more challenging. At its closest approach, Mars is about 50 million km from Earth; at its farthest almost 402 million km (when Earth and Mars are on opposite sides of the Sun). The average distance between Earth and Mars is over 200 times as far as it is from the Earth to the Moon [6]. Distance between the earth and moon is about 384,400 km. These far of regions form the earth where the planets reside are known as deep space regions.[1] For many years the USA and Russia are showing significant interest in this region and are found sending interplanetary probes and undertaking research. Some Asian powers, namely Japan, China and India, are also found showing considerable amount of interest in this region for last few years. These states have already successfully completed their Moon missions. Japan and China have also attempted Mars mission though not successfully. It is important to note that for last four decades (post Apollo 17 mission/December 1972), no human supported activity has

---

[1] Any satellite (or a probe) which travels to a distance of 100,000 km or more from the earth's surface is known to have entered the region which is normally depicted as deep space. This is the region beyond the gravitational influence of Earth encompassing interplanetary, interstellar and intergalactic space. As per astronomical definition, this region is any region of outer space beyond the system of the Earth and Moon. However, when put in the context of distance, the Earth's Moon could be viewed to be in the deep space. Refer [7]

happened in the deep space. In fact, the nature of investments, experimentation and expertise in regard to undertaking any missions in the deep space area appears to have ebbed during last few years. Presently, no state including the USA which once had undertaken successful manned Moon missions is capable of undertaking human deep space missions. Among the space-faring states, there appears to be different approaches in regard to venturing into the deep space. Presently, the USA and China are appearing to be more interested to undertaking human deep space missions, while others like India are found keen to invest more in the robotic technology. In regard to human destination to deep space, different priorities do exist. Since the United States astronauts have already conquered the Moon, they are keen to reach the Mars now while China has a plan to put the Chinese human on the Moon.

This book attempts to understand the rationale behind the human quest for the Mars exploration. As a comprehensive assessment for this query is undertaken, it is realised that the basic question "Why Mars?" seeks various responses from technological, economic and geopolitical to strategic perspectives. People from various walks of life could have different opinions about this subject. But, one notion appears to be common and that is the curiosity to know what is out there!

This work mainly concentrates on the forthcoming Mars mission by India. In the process, it also undertakes some implicit questioning of Mars programmes of various other states essentially to facilitate the setting up of the context for a wide-ranging assessment. It also attempts to provide some knowledge on planet Mars and discusses few of the interesting experiments which are been undertaken by various agencies to learn more about Mars.

The number of activities undertaken by various states post 1957 in the outer space could be viewed as the tactful technique of image building by some of them by demonstrating their technological achievements in this largely unknown domain of universe. It also could be viewed as means of developing a sense of identity among its population. Nevertheless, all such benefits could be viewed as peripheral benefits in today's context. Presently, nation-states are found investing in space technologies for the purposes of using this technology in the overall process of nation building in general and socio-economic development in particular.

Different space assets during and after the end of Cold War era have essentially proved their utility for the purposes of remote sensing, meteorology, communication and navigation. All such investments in space have social, educational and economic biases. Essentially, they have a direct "utility" for the mankind. However, in case of other programmes like human space walks, building of space stations for human stay, robotic or manned missions to moon or other planets the efficacy of significant financial and technological investments are difficult to visualise and appreciate. Hence, there are certain amounts of misperceptions in this regard particular from the common people.

Mars is a part of the galaxy which we live in and there are many other galaxies too, in the universe. For centuries, humans are trying to find answers to various questions in regard to the universe as such. As the science and observational capabilities have improved over the years, some earlier perceptions above the planetary activities are also found changing. Few century's back, two important scientific theories were propounded debating about the centre of the universe. One, the geocentric model or geocentrism argues that the Earth is at the orbital centre of all celestial bodies, while the other, the heliocentric model or heliocentricism offers an astronomical model in which the earth and planets revolve around a relatively stationary Sun at the centre of the Solar System. The second model was given by the famous astronomer Nicolaus Copernicus and was published around 1530s/1540s period. His ideas contradicted with the understanding of the Bible then. In the heliocentric model, Mercury, Venus, the Earth, Mars, Jupiter and Saturn all revolve around the sun. The Moon is the only celestial sphere in this system which revolves around the earth, and, together with it, around the sun [8]. In this system, Mars is the fourth planet from the Sun and is visible through the naked eye. Human interest in Mars has been since then. There have been some historical evidences [9] of early telescopic observations of Mars. Various missions to discover the romance of Mars have been taken on so far, and it has helped to present a better understanding of Mars. But, on the whole, Mars missions are found posing a greater challenge to scientific community, and there have been more disappointments than successes. Nevertheless, humans are not ready to accept such failures, and hence, there appears to be a new vigour for going back to Mars in the twenty-first century.

Mars is one planet which has always been described as the most suitable planet for future human colonisation. Nonetheless, the present human knowledge of Mars and status of technology of reaching Mars appears to be wanting. Hence, there is a need for the exploration of Mars from different perspectives. Knowledge of the geophysical, climatic and various other aspects of Mars are essential in order to "conquer" the planet eventually.

This book has four parts. The first part is more introductory in nature and highlights what planet Mars is all about. It also presents a case that why states are keen in investing in activities about learning more about Mars. Second part presents the India case giving specific details about the India's proposed (November 2013) Mars mission. This part also discusses the interests of Japan and China in this regard. The third part provides a narrative about the past, present and expected investments by various agencies in deciphering Mars. Also, an argument has been presented debunking the perception that missions like Mars are wastage of money. This part end with a proposition about the need for a joint mission for Mars by various global space agencies and with a suggestion that how India could take a lead to make such idea work. Last part of the book mainly presents interviews of scientists directly involved in designing, developing and implementing India's Mars mission and also of few other observers.

This work is an attempt to set the basis for recognizing the geopolitical, technological and economical capability of the states to undertake deep space missions such as Mars Mission. Nation-states understand that fulfilling the dream of "Boots on Mars" for any country would raise its international stature significantly. There also exists a very realistic possibility that because of the financial and technological challenges various states could be keen to undertake joint Mars missions.

The research carried out for this work is based on open-source information and based on some interviews and the information shared by the Indian Space Research Organisation (ISRO) and author's visit to few ISRO facilities involved in Mars mission related work. Efforts have been taken to ensure the inclusion of factually accurate information, recognizing the fact in some cases at times, multiple sources are found presenting varying nature of information.

References

1. Pike J, Stambler E. Space Power Interests. In: Hayes P, Westview Press, Oxford. 1996; p.31.
2. A task group report titled space science in the 21st Century, National Academy Press, Washington. 1998; pp. 1–72.
3. Launius RD. Frontiers of Space Exploration. London: Greenwood Press; 1998. p. 29–38.
4. http://io9.com/5024619/aldrin-blames-lack-of-interest-in-space-program-on-science-fiction, accessed on Apr 20, 2013.
5. Connor S. Desperately seeking life on Mars. Space Policy. 2002;18:267–269.
6. http://www.digipac.ca/chemical/mtom/contents/chapter1/marsfacts.htm, accessed on Dec 15, 2012.
7. http://www.thefreedictionary.com/deep+space, accessed on Dec 20, 2012.
8. The Copernican model of the planetary system—the heliocentric model, http://muse.tau.ac.il/museum/galileo/heliocentric.html, accessed on Oct 25, 2012.
9. For various historical anecdotes may refer to William Sheehan, The planet mars: A history of observation and discovery, 1996.

Chapter 2
Why Mars?

Mars is there, waiting to be reached.

Buzz Aldrin

The process of policy making is expected to be a rational process. Usually, the process involves a clear identification of the problem/issue at hand followed by overall assessment of the problem and recognition of the probable options and finally suggesting specific time-bound solutions. However, when the problems are complex, solutions are hard to come around and that at times even there is no clarity about the desired outputs, then the process of policy planning becomes extremely complicated. In many cases, technological and financial limitations restricts the process of policy planning. The process of policy planning could also be "political" where views, ideas, perceptions, lobbying, interest groups and way of negotiations play a vital role towards reaching a definitive conclusion.

Policy making in outer space arena has always been a challenge for both developed and developing states. Although the multiple objectives achieved by the satellite technologies are well known, still various concerns are found be raised with respect to new and innovative missions that are proposed to be undertaken in the space. This happens mainly because of the significant financial investments required for conceptualising new areas. At times, the sheer enormity of the project makes people disbelieve that the proposed idea is workable. Also, different analysts and policy makers view a range of ideas differently and offer their suggestions in regard to immediate priorities for the state. From a politician's point of view whose period in power usually lasts for 4–5 years to make investments into projects which have gestation period of 10–15 years becomes difficult. All this further complicates the process of policy making in the space arena. Particularly for exotic and costly missions like humans visiting space or planetary missions where the mission objectives are somewhat ambiguous at this stage, various uncertainties do remain and policy decisions could always be challenged.

In respect of deep space missions, the missions to other planets about whom we know little and reaching there is both technologically and financially challenging

---

*Buzz Aldrin was the second person to walk on the Moon, quoted in http://www.jamiefoster science.com/education/podcasts/students/Samantha_Carroll/visit2mars/SamanthaCarrollVisit 2Mars.pdf, accessed on Apr 2, 2013.

A. Lele, *Mission Mars*, SpringerBriefs in Applied Sciences and Technology, DOI: 10.1007/978-81-322-1521-9_2, © The Author(s) 2014

such queries are always posed. Hence, the questions like "Why Mars?" and "Why Mars Now?" are found being raised in recent times since few states are planning missions to Mars. Perhaps such type of questions has no straightforward answers and could be argued either way. For a nation-state, there could be multiple reasons for undertaking such challenging and expensive projects. Also, it may not be necessary for every state undertaking such missions to have similar reasons. States could plan and invest in such missions for political, strategic, scientific, social and/or economical reasons. Hence, in order to appreciate the process of policy planning for undertaking such missions, it is important to carry out assessments at different levels of motivations for the states.

Every technology has its own importance and role. It may be incorrect to grade/value the technologies based on their direct scientific benefits alone. The "management" of technology involves reasons other than scientific. Strategic and commercial importance of a particular technology forces the states to invest more towards the development of that technology. Hence, few specific technologies could gain more prominence over others. The most obvious example of this is the nuclear technology. In regard to space technologies, similar trends are also visible. Nuclear and space technologies find greater relevance over other technologies essentially because of their strategic relevance. The application of nuclear and space technologies for peaceful purposes is actually far greater than its military utility. However, these technologies play a significantly important role in regard to determining the deterrence potential (includes nuclear weapons, delivery systems and reconnaissance assets) of a state. These technologies have significant political importance and there exists certain amount of ambivalence with respect of policy formulation in regard to usage and transfer of these technologies. Presently, various states are still in the phase of "experimenting" in the deep space arena and have less clarity in regard to the exact likely strategic gains from such investments. Hence, country's policies for investments into missions like Mars mission needs to be viewed at the backdrop of the relevance of space technologies in the overall global order.

Global order is essentially a contested concept. It could be viewed as the political, economic or social situation in the world at a particular time and its effect on relationships between different countries [1]. It is difficult to offer an exact definition of the global order mainly because order is mostly viewed as dynamic idea and as a matter of degree. In various cases, at the conceptual level, there are always competing ideas of what constitutes national interests, desirable foreign policy goals and connected views of global order. Probably, where you stand on the matters of order also depends on where you sit in the global hierarchy. The global nuclear regime (non-proliferation treaty, NPT and other related treaty mechanisms) could be viewed to have played an important role towards the maintaining of global order [2]. However, in the Cold War era and post-Cold War era, few states are found using nuclear non-proliferation regime as a means to restrict access to few other states to the nuclear club. This also could be viewed as an attempt to restrict the global order. The question is, "would space technologies also meet a similar fate?" Probably, it is too premature to answer this question at this stage but such issues could gain importance in years to come when humans would have achieved far more success with their space endeavours.

For effective maintenance of international order, it is also important to address various global challenges from nuclear proliferation, climate change, disaster management to terrorism collectively. In order to harmonise various state and non-state actors to accept the notion of global governance over the years, various processes of norms formation have been undertaken. Mostly the norms formation, international law, sustainable development and international relations, etc., could be viewed as constituents in respect of global order. The notion of globalisation which is essentially about the integration of economies of the world could also have its say in regard to global order.

Technology plays an important role in global order. Technology and globalisation are mostly mutually dependent. Since the beginning of the first industrial revolution in the eighteenth century, the technologies could be viewed to have played a greater role towards maintenance of the global order. The process of technological change spurs structural changes in the economy and society [3]. In a technology-driven global order, technological capabilities of a country are an important determinant of its competitive strength both at home and in the world market [4]. Nuclear technologies, space technologies and the military industrial complex have played an important role in the dynamics of the overall global order. America's success with the Moon mission in 1960s was viewed as a "victory" over the erstwhile USSR in Cold War era. In the present era beneath the changed geopolitical settings, it is important to situate the reasons for the investments in Mars programme in the broader context of global order.

Every nation-state intends to design their space policy as a subset of their overall national agenda. It is expected that the civil, commercial and military space policy would be designed to meet the practical demands and requirements of the state and mostly states are found catering for economic realities before making such investments. However, many a times the "motives" of the actors associated with the space policy making are also found dictating the future course of policy making. Individual and organisational interests also impact the process of policy making. At times, this leads to the evolution of space programmes and projects those may not be well optimised, costly and time consuming. Space policy has to accommodate a broad range of perceptions and interests, from practical issues of national defence, commerce and technology to less quantifiable characteristics such as the contribution of space exploration and development for societal benefits and to the achievement of humanity as a space-faring species [5].

The intellect and consciousness of human mind are always urging it to inquire more about the universe than what is known. In fact, the human fascination with the universe has been there for long and would always remain. For many years, scientists are working on various aspects of the Big Bang theory a prevailing cosmological model that describes the early explosive origin of the Universe. Humans have always kept their interest alive with regard to finding out about the presence of life on other planets or other solar systems. There is an interest to know about the birth of stars and planets. The field of the study of celestial objects commonly known as Astronomy has attracted the attention of many inquisitive and brilliant minds. It could be traced back from the Galileo (1564–1642) period when he first looked at sky by constructing his own spyglass and then went on to construct the telescope. Since then, significant developments have taken place in the field of space sciences which essentially involves studying various known and unknown aspects of the outer space. Mainly the issues related to astronomy, planetary sciences and cosmology do get discussed under space sciences.

Human endeavour for space exploration could be for various reasons. Modern day space programmes could be categorised under two basic groupings. One, the agenda driven for knowing more about the universe, galaxies, unidentified matter, to reaching the planets in our solar system; and other is the agenda driven by developing technologies for launching satellites in low, medium and geostationary orbits and building space stations and carrying out experiments in the zero gravity atmosphere, undertaking robotic/human missions in low Earth orbits. One aspect of the space exploration could be essentially identified with the radars, astronomical outposts and ground- and space-based telescopes and theorists developing computer-based simulation models to know more about the galaxy formation, while the other facet of space exploration is about the satellites, spacecrafts, space stations and astronauts. It is important to note that both these groupings are considerably interdependent.

Post Sputnik, there has been a significant propaganda element associated with the exploration of the universe. In the Cold War era, reaching the Moon had more political messaging than actually gaining any knowledge about the Moon. But, the same may not be totally true in the twenty-first century. Probably, now various states are undertaking cost-benefit analysis before making major investments in the space exploration related projects. But, it is important to note that along with the economic costs, the states are also factoring in for the political and technological costs of their investments in space arena.

It is difficult to actually identify the exact reasons for the state to undertake mission to Mars. This is not to say that the states do not build up a perspective before undertaking such missions but the current endeavour is to know more about Mars and based on this knowledge decide the future course of action. A mission to Mars is also a subset of the overall space agenda of the state. Hence, investments into Mars could be viewed as a continuum of state's policies. Space-faring nations do plan to have their own "dream" missions in space and Mars is just a step in that journey. Planet Mars has some peculiar attractions for which the states are dreaming to reach there. In general, the quest for the red planet emerges out of its own logic.

Post the launch of Sputnik satellite by the erstwhile USSR during 1957 broadly, three priorities [6] for the development of space programmes by the states were identified:

1. Cold War era rivalries and desire to demonstrate the technological superiority
2. Lure of discovering unknown
3. Adventure

Presently, apart from the Cold War era rivalries, the other reasons mentioned above are still valid and could be viewed as motivations for the states to dream of Mars. Today, even in the post-Cold War era, the United States would not like to "loose" its technological leadership because it plays an important role in maintaining their superpower status. They are not oblivious to the fact about China's attempts to make major inroads in the technology domain. The USA understands that China is using their space capabilities to demonstrate their technological proficiency. With China displaying the traits of emerging superpower, the USA is unlikely to surrender their leadership in space arena. Hence, Mars missions including the human Mars missions do have their own geopolitical insignificance.

Our solar system consists of the Sun and eight official planets. The inner solar system contains the Sun, Mercury, Venus, Earth and Mars. While the outer solar system planets are Jupiter, Saturn, Uranus and Neptune, while Pluto has now been classified as a dwarf planet during August 2006 (by the International Astronomical Union). For various reasons, it is been believed that apart from the Earth, probably the Mars could be the most habitable planet in our solar system. There are varieties of reasons for such beliefs. The length day over Mars is similar to Earth, around 24 h. Also, its gravity is 38 % that of Earth's, which could be manageable for human survival. The ground temperatures over Mars range around 0–100 °C. It has an atmosphere which could provide protection from cosmic and the Sun's radiation [7]. No other planet offers such manageable similarities with the Earth for a human stay. Also, Mars has its two Moons namely Phobos and Deimos and understanding more about them could also help to know more about "a planet and a Moon system", in general. This could help to draw some inferences with regard to Earth–Moon system.

Human beings are trying to find an answer to a basic query that why Earth is such a beautiful place to live and would it always remain like that? Why the Earth is just right distance from the Sun—not too far for the oceans to freeze or not too close for the oceans to boil? Planetary scientists call it "Goldilocks Paradox". It is argued that Earth and its two neighbouring planets Mars and Venus were formed around the same time about 4–6 billion years ago, from the same ingredients including water, carbon dioxide and nitrogen. But, only Earth developed life. Mars is too cold and Venus is too hot, but Earth is just right [8].

However, humans are yet to find the answer for such paradox. We are yet to find a theoretical backing for why do the laws of physics seem fine-tuned for life? The Mars exploration could help us to understand about the evolution of Earth. Probably, the study of other planet which has a matching DNA with Earth could help us to understand more about the origins of volcanoes, earthquakes and weather [9]. All such studies over a period of time could directly or indirectly benefit humans to address issues related to global warming/climate change and also forecasting of probable natural disasters.

It is important to note that climate change happens not only because of the factors originating on the Earth itself like human-specific impacts on the surroundings, but it could also occur because of the external factors like solar radiation received by the planet, etc. Also, scientists are trying to understand how the variations in the Earth's climate by the changes in the characteristics of the Earth's orbit and axial tilt could take place. Study of current weather changes on Mars with changes in its atmospheric composition both due to on planet and outside planet factors could help to develop atmospheric models to understand changes in Earth weather [10].

Study of the Moons of Mars could help to know more about the asteroids in particular and the formation of the solar system in general. The study of Phobos and Deimos will tell us about the structural strength of asteroids and these Moons are also excellent sites for the study of impact processes. Exactly how craters come to look like craters is still a subject of some debate and Mars missions could help to find answers to this. As per one school of thought, crater morphology depends on the body's surface gravity while some other studies conclude that material strength and the impact velocity are of more importance. Maybe a Mars mission could allow us to reach to a singular conclusion. In addition, some first-hand knowledge could be gained in regard to how craters are created in a very low gravity environment. All this would help in better understanding of Earth [11]. There are many unanswered questions like why Earth is the only water-rich planet, was there water available on other planets too and if so, why did it disappear? Some answers to such question could help us to know the future of water on the Earth. Also, mission to Mars could assist in knowing more in various arenas form the planetary geophysics, from the knowledge about the role of carbon dioxide in the atmosphere to the evidence in regard to the life on Mars. It is known that Sun has a direct correlation with the weather on the Earth. However, because of the particular rotation of Earth and Sun at times, it is not possible to study all the properties of Sun form Earth. Such study could become possible from Mars.

Apart from various scientific benefits undertaking of Mars mission is expected to offer various technological benefits too. From Launcher to Lander, a range of technologies would be required to make the mission happen. Every new launch to the Mars would demand the development of additional technologies. Also, the complex nature of such missions would force the development of some innovative technologies.

Specific propulsion technologies would be required to provide the energy to get to Mars and conduct long-term studies. Entry, descent and landing technologies would be required to ensure precise and safe landings [12]. Nature of mission (human or robotic) would also dictate the requirement of additional technologies. There would be a requirement to invest into a range of different technologies from avionics to planetary protection technologies to remote sensing sensors to tele-communication technologies. Human travel and survival would necessitate investments into an array of new technologies. All this could lead to major development to the field of robotics, control systems, materials, communications, nano- and biotechnologies, artificial and ambient intelligence, etc. Subsequently, over a period of time, there are chances that various spinoff technologies could get developed in the fields of transportation, information and communication technologies, robotics, medicine, agriculture, etc.

No political leadership is expected to take decision in vacuum with regard to their state making investments in programmes like Mars mission. There has to be a justifiable case put forth by the scientific community in this regards explaining both the need for such a mission and their capabilities to successfully pull it through. Designing the right mission at the right budget is obviously a requirement for going forward [13]. At the same time, it is also important to note that a cost-benefit analysis purely based on economical viabilities may not be prudent to decide the fate of various new space missions. This is mainly because such investments may not look viable in short term but are likely to offer benefits in long term. Also, since various missions undertaken in space have direct or indirect benefits for society on range of activities from education to science research to disaster management, non-commercial funding for space is mandatory. Present phase is a period of Mars exploration; hence, the financial costs incurred should be viewed as an "investment".

Political leadership is expected to play a very important role for a state to invest into programmes like Mars mission. A leader of celebrated vision and conviction could only appreciate the importance of such agendas for their state. Here, again the presence of a reputable technocrat, presenting a credible technology leadership is also vital. Presently, the financial crisis witnessed in the twenty-first century is restricting the space-faring nations to plan for gigantic space agendas. However, it is felt that a vigorous, focused, goal-oriented space programme would eventually bring more employment opportunities, assist industrial growth and open the spigots of technological innovation.

It is also felt that even though the investment would eventually come if is felt that there could be a reluctance to invest in human Mars programme. The present thinking appears to be more biased towards undertaking robotic missions. The perceived lack of interest in human space exploration could be because of two main reasons: one a fear of human casualties, and second a misguided belief that we must solve all our terrestrial problems before doing anything ambitious in space! [14] However, it is important for policy makers to appreciate that attempting to resolve only extremely difficult scientific challenges would help us advance further and Mars mission offers such options.

The question is "why Mars now?" A simple argument could be since more than 50 years have passed as humans have succeeded in reaching space if not now then when? Humans could be said to have started trying to understand the secrets of Mars since 1600s with the invention of telescope. In the space era, attempts have been made since early 1960s to study Mars by sending probes in the vicinity of Mars. This was happening as a part of the unmanned spacecraft interplanetary exploration programme undertaken by the erstwhile USSR and the US.

In early years, Mars was found unfriendly to Earth's attempts to visit it. More missions have been attempted to Mars than to any other place in our Solar System (except the Moon), and almost 50 % attempts have failed. Various initial failures could have happened probably because Mars was the first planet Earth attempted to explore [15]. These failures have also taught us many lessons and assisted in making few subsequent missions more successful. But, still space powers are yet to master the art of reaching Mars and some disappointments have occurred relatively recently. Luckily, some successful missions since 1996 have provided important data about Mars helping us to better understanding. This is helping a better planning for future missions.

One of the questions with regard to quest for Mars is that "are humans aiming for the mineral deposits on the Mars?" However, it is still premature to answer this question. No definitive information in regard to the Mars mineralogy is available. However, some studies are available providing the regional surface material distributions on Mars [16]. There are indications that the soil on the Mars surface could have volcanic origin. Also, clay minerals that usually form when water is present for long periods of time covering a larger portion of Mars than previously thought [17]. Mars has a different crust than Earth, and very different atmosphere and so the minerals over there are expected to be different than that of Earth [18]. Based on available information mapping of the Mars surface has been done by scientists which give a reasonable idea above the area for further studies and excavation from the mineralogy point of view. However, even if some useful minerals are found over there is it highly unlikely that they could be transported back to the Earth[1]. The existing technology cycle even at its maturity peak would not offer effective, viable and economical solutions to transport back any material from Earth to Mars.

Another aspect of Mars which states are likely to consider in their assessment particularly from the point view of human missions to Mars is the climatic conditions over the Mars. Even though in comparative sense probably Mars could be the most habitable planet in our solar system after the Earth, still the issues related gravity, low temperatures and absence of atmosphere offers various challenges for a long human stay. Bulky equipments would be required to be carried for human sustenance posing various technological and economical challenges. All this to a greater extent could discourage the policy makers.

Also, there is an opinion that benefits projected from Mars mission are "vague" in nature and more clarity is required on actually what could be achieved from such missions. Hence, a policy maker could raise questions that "Is Mars worth the risk?" and "Worth the financial investments?" It is important to note that the space travel is analogue of national esteem and a momentous act like human visiting Mars is bound to bestow a great power status to the country. Hence, any myopic economic approach is not advocated. Also, the success of few recent missions to Mars particularly the success of the United States rover Curiosity should provide encouragement for the further research.

The argument like "there is a lack of real societal support for such costly missions" needs to be assessed appropriately. It needs to be appreciated that opinion polls conducted in regard to the efficacy of Mars mission have limited shelf value and do not always represent the view of majority. Moreover ill informed

---

[1] Ghosh A. A scientist working on NASA's Mars programme has expressed this opinion while speaking with the author.

media debates on such issue vitiate the public opinion. On the other hand, success of missions such as Curiosity brings change in opinion of many people. People want their scientists to achieve stupendous successes with their missions in space. Also, scientific community is always keen to take calculated risks with their research and if they achieve a significant success then do get major public support for future missions.

Whether life has existed on Mars or not is still an open question. There is a need to undertake detailed biological experimentation including study of fossils in this regard. Humans are keen to get the answers with respect to presence of life outside Earth and are ready to make all efforts towards finding an answer to this. The nature of technological developments witnessed in space area bestows the confidence that albeit the human mission to Mars may be a difficult proposal but definitely not an impossible idea. Particularly, in the twenty-first century with few states already having acquired some amount of a success in knowing more about Mars, it is worth pushing the envelope further up. Overall conquering the Red Planet appears to be worth the risk.

It is obvious that the states would decide their Mars agenda based on their technological preparations once the policy decision to undertake a mission has been taken. However, there are only specific launch windows which are available when planning a launch mission is advisable. For an interplanetary launch, the window is constrained typically within a number of weeks by the location of Earth in its orbit around the Sun, in order to permit the vehicle to use Earth's orbital motion for its trajectory, while timing it to arrive at its destination when the target planet is in position. Every 26 months, Earth, Mars and the Sun align for the most efficient, least energy-consuming path between Earth and Mars [19]. Actual planning of any Mars mission would depend on the planetary alignments. States have to manage with such limited window availability. The present-day rocket technology allows the crafts launched from the Earth to reach Mars within 8–9 months. Scientists are working on various technological aspects with an aim to reduce this period significantly. For any future human mission to Mars, the one-way travel time of nine months to reach the Mars orbit is not particle.

In post-Cold War era, there have been always talks of space race particularly in Asian context and that too mainly involving the states like China and India. The reasons for this are obvious and mostly geopolitical. Such talks become loud particularly when both these states start planning for similar missions. Naturally, since both these states have interests in Mars, the issue of space race is getting discussed at various forums and particularly in media where there are many takers for such notion. It is important to appreciate that when comparison is carried out in respect of technology and budget investments, China is much ahead of India in space arena. Also, India is unlikely to view every Chinese investment in space as an act which needs to be responded suitably. In few cases, their programmes do have some commonalities and when such investments are scrutinised at the backdrop of geopolitical veracities, it is possible to (vaguely) conclude about the possibility to race. In an interview with the author Joan Johnson-Freese[2] has following to say: "I am always sceptical about media reports on 'space race' activities for two reasons. First, very often reporters are not accurate in their reporting, and second, plans are only important if there is a budget attached. Very often I see statements about China's long term goal of putting a man on the Moon, when in fact a manned lunar mission has not been approved in China and in fact has only recently been a topic of discussion. The ultimate goal of their three-step plan being implemented since the 1990s is a large space station. China has reaped significantly geostrategic benefits from their many achievements and 'firsts' as part of that program. But what comes next remains to be seen. They have launched one probe to Mars, Yinghuo-1, which failed. Clearly China would like to expand its interplanetary missions, but how rapidly and how aggressively remains to be seen. Yet clearly much of what is being done in space right now in Asia is for geo-strategic reasons. It appears to me that India's plans to go to Mars with the Mangalyaan mission positions India to leap ahead of China in this one area—being only the fourth country/organization to reach Mars (with the United States, the then Soviet Union, and ESA)—ahead of China—whereas India has been behind China in other key areas of space exploration and development. What I will be watching in the future is which country is willing to put the required budget behind their plans to show not just the success of one mission which looks good in the record books, but a sustained programme of technology development and launches".

Based on the overall debate in regard to various aspects of Mars agenda, the reasons for various states opting for the Mars mission could be identified as (1) fascination for the Red Planet, to know what it there, if life ever existed (2) technological challenge, humans have reached Moon so Mars is next logical step (3) to boost studies in space technologies (also spinoff technologies could offer additional benefits) and planetary sciences (4) nationalism, great power status and display of strategic superiority (5) economic advantages/boost to space industry.

References

1. http://www.macmillandictionary.com/dictionary/british/world-order, accessed on Dec 20, 2012.
2. Foot R, Walter A. China, the United States and Global Order, Cambridge University Press, Cambridge; 2011.
3. Technology and Global Order http://www.it.iitb.ac.in/~prathabk/pages/tech_archives/global/technology_and_global_order.pdf, accessed on Dec 24, 2012.
4. http://sts.sagepub.com/content/6/1/23.abstract, accessed on Dec 18, 2012.
5. Sadeh E, editor. Politics of space. London: Routledge; 2011. p. 3.
6. Launius RD. Frontiers of Space Exploration. London: Greenwood Press; 1998. p. 5–6.

---

[2] She is professor at the naval war college, USA and author of several books on space security.

7. Why Mars? Why not another planet?, http://mars-one.com/en/faq-en/22-faq-mission-features/199-why-mars-why-not-another-planet, accessed on Dec 12, 2012.

8. Anderson I. Model atmospheres show signs of life, New Scientist, Jan 7, 1988, p. 41.

9. http://www.astrodigital.org/mars/whymars.html, accessed on Dec 10, 2012.

10. http://www.astrodigital.org/mars/whymars.html, accessed on Dec 10, 2012.

11. http://www.astrodigital.org/mars/whymars.html, accessed on Dec 10, 2012.

12. http://marsrover.nasa.gov/technology/, accessed on Dec 24, 2012.

13. Why Mars matters, http://articles.latimes.com/2012/aug/05/opinion/la-ed-mars-curiosity-nasa-budget-20120805, Aug 5, 2012, accessed on Sep 16, 2012.

14. http://spectrum.ieee.org/aerospace/space-flight/why-mars-why-now/0, accessed on Oct 26, 2012.

15. http://www.planetary.org/explore/space-topics/space-missions/missions-to-mars.html, accessed on Nov 27, 2012.

16. Bandfield JL. Global mineral distributions on Mars, J Geophys Res. 107(E6); 2002. pp. 9–19.

17. Clay minerals abundant on Mars than thought, Deccan Herald. Dec 21, 2012.

18. O'Hanlon L, Mining mars? Where's the ore? http://news.discovery.com/space/mars-prospecting-ores-gold.html, Feb 22, 2010, accessed on Dec 31, 2012.

19. http://www2.jpl.nasa.gov/basics/bsf14-1.php, accessed on Dec 20, 2012.

Chapter 3
Discerning Mars

> For Mars alone enables us to penetrate the secrets of
> astronomy which otherwise would remain forever hidden from
> us.
>
> Johannes Kepler

3.1 Natural Features and Atmospheric Conduction

Interests in Mars are almost universal. The romance of the decadal old idea of
human settlements over Mars has not yet died down. Probably, every space-faring
state has dream of reaching Mars. Some of them have announced it openly and
found making efforts to achieve this and perhaps the others due to technological
and financial constraints are following wait and watch approach.

Mars is the fourth planet from the Sun and the second smallest planet (Mercury
is the smallest) in our Solar System. The planet is recognised by the name "Red
Planet". The reddish appearance of the planet because of the iron oxide present on
its surface is the basis for this name. The name Mars has its origins in the Roman
era in which Mars was recognised as a god of war (Fig. 3.1).

This planet is about half (53 %) the size of Earth, but because Mars is a desert
planet, it has the same amount of dry land as Earth. Mars is not a sphere and its
shape is called an oblate spheroid. At its equator, Mars has a diameter of 6,794 km
and circumference is 21,343 km. From pole to pole, the diameter is 6,752 km and
circumference is 21,244 km around [1].

Mars is the place for highest mountains and deepest valleys. One of the
mountains (called Olympus Mons) over Mars is about three times in height than
the Mount Everest. This mount is the large shield volcano, a volcano made mostly
of fluid lava flows (however, Mars is no longer volcanically active) [2]. In spite of
being thinner, the Mars atmosphere is still capable of supporting weather. Clouds
and wind are the main weather elements observed. Dust devils and dust storms are
experienced by the Martian surface. Wind is the main force that sculpts Mars
features. Giant sand storms regularly scour the entire planet. At times, it even
snows on Mars and the snowflakes are made of carbon dioxide.

---

*Author of the New Astronomy. 1609. http://starryskies.com/The_sky/events/mars/opposition
01.html. Accessed on Apr 19, 2013.

A. Lele, *Mission Mars*, SpringerBriefs in Applied Sciences and Technology,
DOI: 10.1007/978-81-322-1521-9_3, © The Author(s) 2014

Fig. 3.1 Solar system [23]

Today, the Martian surface preserves the greatest range of elevation on any of the terrestrial objects, from a 1,300 km-wide crater in the northern hemisphere, whose floor is 10 km below mean elevation, to the 26 km-high Olympus Mons, the largest volcano in the solar system. Olympus Mons is an example of a shield volcano; a volcano that gets formed when thin lava spreads out over a large area (such volcanoes have broad sloping sides and are generally encircled by gently sloping hills in a fan-shaped pattern, which looks more like a warrior's shield). The diameter of this extinct volcano is about 600 km, or three times the wide size of the largest earth volcano of this type (located in Hawaii). Such volcanic resurfacing of the planet has affected primarily the planet's northern hemisphere, which is covered with relatively smooth plains while the southern hemisphere surface shows the accumulation of many more craters over time [3].

There are less than 20 volcanoes on Mars, and scientists have inferred that the driving force of early volcanism could be due to evolution of mantle hotspot. Presently, there are no signs of any Martian geological activity. Last lava flow occurred roughly 20–200 million years ago. As per the 2007 study by National Geographic, the volcanoes may erupt and could spew carbon dioxide and water into the atmosphere. Also, the red planet has polar caps that wax and wanes with the Martian seasons. Seasonal polar caps are made of Martian air that freezes during the winter. These polar caps are made of carbondioxide (the major air component of Martian atmosphere) [4].

One of the major features on the surface of Mars is a long deep valley which runs close the equator. This is a system of canyons called Valles Marineris is a 4,000 km-long, 600 km-wide and 7 km-deep valley. Apart from this, one of the most spectacular features on the Mars is Hellas Planitia, also known as the Hellas Impact Basin. This largest visible impact crater in the solar system is in the southern hemisphere and is of 2,300 km in diameter and few km in depth [5]. While the North Polar Basin, or Borealis basin in the northern hemisphere constitutes almost 40 % of Mars, many craters, especially at high latitudes, surrounded by fluidized ejecta suggest the presence of underground water or ice in early times. Based on examination of Martian meteorites, it was found that surface of Mars is primarily composed of basalt and a large part of the surface is deeply covered by dust of Haematite. NASA's Phoenix lander revealed that Martian soil is slightly alkaline and contain nutrients like sodium, magnesium and potassium. Results from Mars Reconnaissance orbiter and Mars express confirmed the presence of large quantities of water ice at the poles. Sub-surface level remote sensing indicated the presence of large quantities of water trapped underneath Mars cryosphere.

Mars Climate is no more an enigma with various observations been made available for the Martian missions. However, scientists are yet to conclusively establish the theory that the climate on Mars 3.5 billion years ago was similar to that of early Earth: warm and wet. The seasons in one hemisphere (South) are more extreme while in the other (north), they are less extreme. Presently, the Mars atmosphere is very thin, the temperature is very cold. The current climate changes drastically during the year. It has seasons similar to the Earth's due the tilt of its axis. But because its orbit around the Sun is elliptical, the distance from the Sun varies about by 20 % depending on where it is in its annual orbit. Mars is at greater distance from the Sun in comparison to the Earth (average distances 150 million km vs. 228 million km) and this has got the implications for the temperatures on these planets. Also, the low temperature of Mars is a result of the presence of large amount of carbon dioxide, which radiates away the energy that reaches the planet from the Sun [6]. The average Mars temperature is around −60 °C (very at different locations like poles and equators and catering for the seasonal variations, they are around in the range between −125 and +20 °C). Barometric pressure varies at each landing site on a semi-annual basis. The spacecrafts which has reached Mars undertaken pressure observations show the readings as low as 6.8 millibars and as high as 9.0 millibars [7]. Generally, it has been observed that pressures from location to location and also from season to season. Even readings like 10.8 millibars have also been recorded. In comparison, the average pressure of the Earth is 1,000 millibars.

Mars has a very thin atmosphere about 100 times thinner than Earth's and incapable of easily supporting life. The composition of its atmosphere is as follows [8]:

- Carbon dioxide: 95.32 %.
- Nitrogen: 2.7 %.
- Argon: 1.6 %.
- Oxygen: 0.13 %.
- Carbon monoxide: 0.08 %
- Also, minor amounts of water, nitrogen oxide, neon, hydrogen–deuterium-oxygen, krypton and xenon.

As seen above, the Chemical composition of the Martian atmosphere is rather simple: CO_2, H_2O, N_2 and their dissociation products and noble gases. The observed O_2 and CO abundances are much smaller than the predictions of early dry models and this problem still remains unsolved. Martian air contains only about 1/1,000 as much water as our air, but even this small amount can condense out, forming clouds that ride high in the atmosphere or swirl around the slopes of towering volcanoes. There is evidence that in the past, a denser Martian atmosphere may have allowed water to flow on the planet. Physical features closely resembling shorelines, gorges, riverbeds and islands suggest that great rivers once marked the planet. On Mars, the air is saturated (100 % humidity) at night, but under saturated during the day. This is because of the huge temperature difference between day and night [9].

Carbon dioxide, the major constituent of the Martian atmosphere, freezes out to form an immense polar cap, alternately at each pole. The carbon dioxide forms a great cover of snow and then evaporates again with the coming of spring in each hemisphere [10].

Occasionally, strong winds on Mars create dust storms. After such storms, it can be months before all of the dust settles. These clouds of dust, sometimes raised to 80 km levels in the atmosphere, may be local (up to 100 km region or so), regional (up to 1000 km region or so), or fully global in size and area affected. These phenomena range in size from dust devils (metres in diameter and highly localised) to hemispheric events that can on rare occasions contribute enough dust to obscure the entire planet. Dust storms, especially those of global dimensions, do not occur every Mars year and does not occur with any regularity. Local-scale dust storms can occur during all seasons and are present in both hemispheres. Dust storms have a preference for occurring during the strong solar heating period during southern hemisphere summer months. Atmospheric dust on Mars can take months to settle out of the atmosphere and deposit onto horizontal surfaces.

Martian planet has been studied for many years. Various studies have been undertaken based on information gathered during orbiter or flyby missions and landing missions. These studies initially involved gaining knowledge about the surface, crust and interior of the Mars. The United States Geological Survey (USGS) Astrogeology Research Programme has created an image map of the Mars. They have divided Mars into 30 quadrangles and various and geological details like composition, structure, history and physical processes, have been identified for every quadrangle [11]. There are various scientific papers available written in different periods based on the information available then.

In general, Mars is broadly viewed as a terrestrial planet. It is neither geologically stillborn, like Mercury or the Moon, nor so active that most of the geological record has been destroyed, like Venus or the Earth. The prolonged geological evolution of Mars is recorded in the physical and chemical characteristics of its crust. Some observations indicate that the planet once had a magnetic field, much like Earth does today [12]. The bulk of the about 50-km-thick Martian crust formed at around 4.5 Gyr B.P., perhaps from a magma ocean (1 Gyr is 10^9 years, BP stands for before present). This crust is probably a basaltic andesite or andesite and is enriched in incompatible and heat-producing elements (Andesite is an extrusive igneous volcanic rock, intermediate type between basalt and dacite). Some recent research based on satellite images has provided strong evidence for plate tectonics on Mars. It had been thought, until now, that tectonic movements were only present on Earth [13].

3.1.1 Moons of Mars

The two small moons orbiting Mars were discovered in 1877, by Asaph Hall, at the United States Naval Observatory. In fact, before this discovery of the two tiny moons, Phobos (Flight) and Deimos (Fear), during early 1,700s Johannes Kepler had speculated that since the Earth had one moon and Jupiter had four moons known in his time, Mars might have two moons since it orbits between Earth and Jupiter [14] (Fig. 3.2).

These moons whose names have remerged form Greek mythology are not spherical but more like asteroids. Their orbits are very different from Earth's moon [15]:

- Phobos travels from west to east and makes an orbit in only 11 h (roughly 28 km by 23 km by 20 km).
- Deimos travels east to west (30 h orbit) but takes almost 2.7 days to set in the west (approximately 16 km by 12 km by 10 km).

Since they are tidally locked, both moons always show the same side of themselves to Mars. Both these moons are heavily cratered and most likely are the captured asteroids and not objects that formed in conjunction with the formation of the planet. Phobos orbits at a distance of less than 6,000 km from the surface of Mars. Phobos has an orbital period of 7 h and 39 min. Deimos is even smaller than Phobos. Its longest diameter is 16 km and it orbits 23,400 km from the planet's centre and, unlike Phobos, has a stable orbit. Because Phobos has relatively faster orbit, even faster than Mars rotates, it is constantly being drawn closer to the planet. When it gets close enough, it is expected to crash into the planet in less than 100 million years [16] (Table 3.1).

Various missions to Mars during last four to five decades have succeeded in collecting significant amount of information in regards to topography, terrain, atmospheric conductions, aeronomy, chemical composition and mineralogy. This has allowed states to learn more about Mars but at the same time has also raised few questions, and scientists are trying to find answers to them. Future missions have been planed based on the information received from earlier missions with a

Fig. 3.2 Mars and its Moons [24]

Table 3.1 Comparing the features of Earth and Mars. The table is constructed based on the information available at [25]

Planet	Distance from the Sun (astronomical units miles km)	Period of revolution around the Sun	Period of rotation	Mass (kg)	Diameter (miles km)	Apparent size from Earth	Temperature (K range or average)	Number of Moons
Earth	1 AU 93 million miles 149.6 million km	365.26 days	24 h	5.98×10^{24}	7,926 miles 12,756 km	Not applicable	260–310 K	1
Mars	1.524 AU 141.6 million miles 227.9 million km	686.98 Earth days	24.6 Earth hours = 1.026 Earth days	6.42×10^{23}	4,222 miles 6,787 km	4–25 arc seconds	150–310 K	2

view to fill the gaps in existing available knowledge about Mars and also to gather more information from the point of undertaking future robotic and human missions.

3.2 Mars Astronomy

Mars which is the second nearest planet to the Earth and the fourth planet from the Sun could become the first planet visited by humans if the Mars agenda of few space-faring states progress as per the plans within next two to three decades.

Till date for humans to reach Mars even by undertaking unmanned missions has remained challenging. Apart from the technological challenges, the missions to Mars also have limitations because of the limited launching opportunities available. This happens because the distance of Mars from the Earth various significantly because the nature of the orbits of motions of these planets. After about every 2 years and 2 months (26 months), Mars is at the closest to the Earth. Hence, the suitable lunch window[1] is available only after 26 months to plan a mission to Mars form the Earth. Martian Opposition occurs when Mars is opposite the Sun relative to Earth, meaning that a straight line could be drawn through all three with the Earth in the middle. Mars has an opposition about every 26 months. The following diagram (not to scale) explains this arrangement [17]: (Fig. 3.3).

Mars moves in an elliptical path, and its distance from the Sun varies from 206.5 million km at its closest (perihelion [18]) to 249.1 million km at its farthest (aphelion). The planet completes each revolution in about 687 days. In comparison with the Earth which takes 365 days to complete one revolution, the Mars takes almost double the time. Because of the orbits of the two planets, as mentioned earlier, Mars oppositions occur about every 26 months; the accurate average is the synodic period of 779.94 days. The Earth–Mars distance fluctuates between about 56 and 400 million km (35 and 250 million miles), as the two planets swing around their respective orbits.

The best option available to visit Mars could be when the Mars opposition occurs at the same time that Mars is at perihelion, which means Mars, in its orbit, is closest to the Sun. On Aug 27, 2003, the opposition distance of Mars from Earth was the closest it has been for about 60,000 years[2], it was at the distance 34,645,500 miles (55,756,600 km). Similarly, some 79 years ago, in August 1924, also the Mars was found relatively close to the Earth. It is important to appreciate that the orbits of the planets are not completely round, but rather elongated circles called ellipses. About every 15–17 years, Mars is both at opposition and perihelion, and at these times, Mars is closer to Earth than other times when there is just

---

[1] The time period in which a particular mission must be launched.

[2] A news was also making rounds then when it was "predicted" by few that Mars would be visible of the same size of moon on that day, this event is famously called as Mars Hoax.

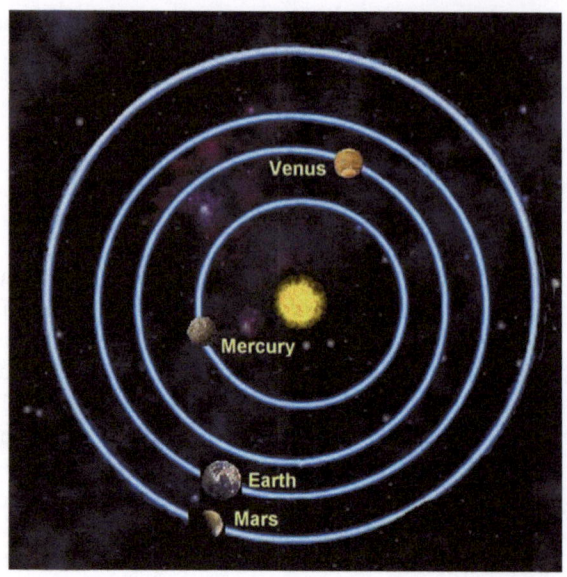

an opposition. In the next millennium, as the minimal distance between the orbits of the Earth and Mars decrease, one will find many oppositions with a Earth–Mars distance closer than that of Aug 27, 2003, but the first such possibility would occur will take place only on Aug 28, 2287, the date on which Mars could come closure than 2003 [19].

If Mars's orbit were nearly circular, similar to Earth, each time Mars has been approached from the Earth it would have be about the same distance from the Sun as usual, and Earth's distance from Mars would be about the same—a little less than 49 million miles. But Mars's orbit is not as circular as Earth, having an eccentricity of 0.093 (compared with 0.017 for the Earth), which means that its distance from the Sun varies by a little over 9 %, or a little over 13 million miles. As a result, depending upon where Mars is in its orbit when we approach it, we can pass as little as 35 million miles from it (if it is near perihelion), or as much as 63 million miles from it (if it is near aphelion).

Simplistically, it could be said that the planets do not follow circular orbits around the Sun; mostly, they follow an elliptical path. Sometimes they are at the closest point to the Sun (perihelion), and other times, they are at the furthest point from the Sun (aphelion). To get the closest point between Earth and Mars, demands a situation where Earth and Mars are located on the same side of the Sun. Furthermore, a situation is required where Earth is at aphelion, at its most distant point from the Sun, and Mars is at perihelion, the closest point to the Sun [20].

As shown in the diagram below [21], the approximately 2 years and 2 months synodic period of Mars also demonstrates that each time Mars would be approached in one synodic period, and every subsequent approach would be a little further along in its (and Earth's) orbit. So if it is near perihelion at one opposition,

Fig. 3.4 Relative position of Earth and Mars during oppositions [27]

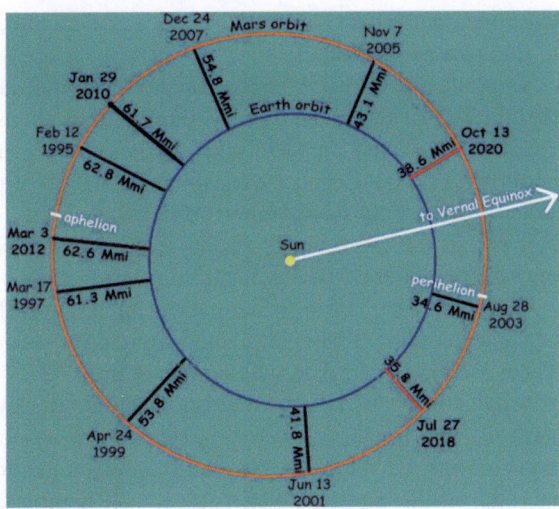

and relatively close to Earth, as it was in 2003, at subsequent oppositions, it will be further and further from perihelion, and further and further from Earth, until it is near aphelion at opposition, and as far from Earth as possible for an opposition, as it was in 2010 and 2012; then, it will be closer and closer to perihelion, and to Earth, at each succeeding opposition, until the next perihelion opposition, which occurs every seven or eight synodic periods, or 15–17 years after the previous perihelion opposition (Fig. 3.4).

Above diagram shows the relative position of the Earth and Mars (in Mmi, or millions of miles) at various oppositions from 1995 to 2012, and the near-perihelion oppositions of 2018 and 2020. Here, the date of opposition corresponds to the Earth's position in its orbit. This diagram is a representation of Earth's relative position and distance from Mars at various oppositions, but only shows one full series of oppositions (all the way around the orbit), and part of another set. The table below lists all oppositions from 1995 to 2020:

- Dec 24, 2007—88.2 million km (54.8 million miles).
- Jan 29, 2010—99.3 million km (61.7 million miles).
- Mar 03, 2012—100.7 million km (62.6 million miles).
- Apr 08, 2014—92.4 million km (57.4 million miles).
- May 22, 2016—75.3 million km (46.8 million miles).
- Jul 27, 2018—57.6 million km (35.8 million miles).
- Oct 13, 2020—62.1 million km (38.6 million miles).

It may be noted that 2018 should be a best year in near future for the Mars mission when it is expected to be close by and particularly bright and red in the sky. It may be noted that when Mars is been attempted to near its perihelion, it takes longer than usual (almost 2 years and 2 months), because it is moving faster in that part of its orbit; and when Mars is being captured near its aphelion, it takes less time than usual (just over 2 years and 1 month), because it is moving slower in that part of its orbit.

Overall, it needs to be appreciated that it's not only the Sun which influences the motions of the planets. Very small perturbations caused by the gravitational effect of the planets on each other also exert some influence. In many cases, such perturbations could cancel each other. But, over a period of time (thousands of years), Mars could have higher orbital eccentricity essentially reducing distance between in the Earth and Mars on some occasions (perihelion opposition). This may happen after 30,000 years from now!

To identify the launch window, it is important realise that planets move in their orbits with different velocities and have relative motion with respect to each other. There are infinite ways in which transfer trajectories between the planets could be generated. But what is important is to meet the energy requirement for establishing most of these transfer trajectories. Hence, it is absolutely essential to find out the transfer orbit which meets the energy constraints. For minimum energy opportunity, the origin and target planets must have appropriate angular relationship at the time of injection of spacecraft into the transfer trajectory. Apart from the energy criteria, there may be other mission constraints like departure conditions, arrival conditions, transfer duration., which reduce the domain of these transfer trajectories.

To determine the range of possible launch dates, it is important to factor in the aspects like the distance between Earth and Mars, the launch vehicle's power, the spacecraft's weight and the desired geometry of approach to Mars, etc. A launch vehicle from Earth should have a design feature to take advantage of the Earth's spin for an added boost, which translate to fuel saving. The point at which a launch vehicle uses the least amount of fuel to push a spacecraft onto the proper trajectory for Mars identifies an ideal launch date. The length of launch window for a mission is often determined by the type of launch vehicle, weight of load and the planet geometry (Fig. 3.5).

Fig. 3.5 Distance of Mars from Earth [28]

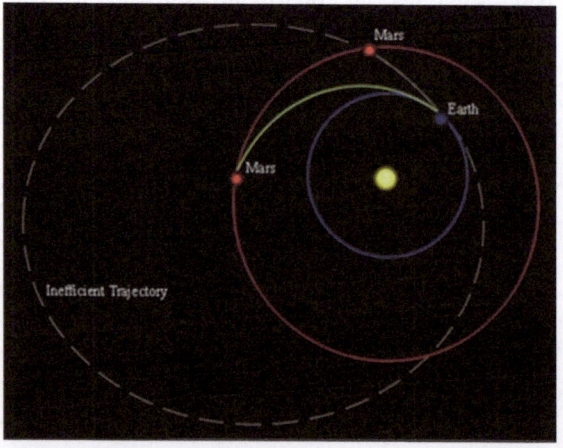

Table 3.2 Approaching launch opportunities for Mars missions. The table is constructed based on the information available at different locations

Earth departure	Transfer days	Mars arrival
29 Nov 2013	314	28 Sep 2014
10 Jan 2016	285	11 Oct 2016
17 May 2018	249	11 Jan 2019

Above diagram indicates that a straight line may not be the best way to reach Mars. This is because the straight line translates into a huge, inefficient orbit around the sun. The launch vehicle required to put the spacecraft in such an extreme solar orbit would have to be very large, very powerful, and very expensive and would waste an inordinate amount of fuel [22].

Since the Earth and Mars orbits are neither exactly circular nor coplanar, for the launch opportunities that exist, the energy requirements are not identical. Minimum energy requirement for various launch opportunities that exists during this decade for placing the spacecraft in an elliptical orbit (500 × 80,000 km) around Mars is as depicted below: (Table 3.2).

It is expected that states have interest in Mars would try to use these opportunities to further their Mars agenda. Particularity, the opportunity in 2018 is expected to be the best option and few private companies are considering this launch window for undertaking the human flyby mission.

References

1. Solar system image available at http://www.seasky.org/solar-system/solar-system.html. Accessed on Apr 2, 2013.
2. http://www.ifa.hawaii.edu/~barnes/ASTR110L_F03/marsmotion.html. Accessed on Mar 1, 2013.
3. http://www.ifa.hawaii.edu/~barnes/ASTR110L_F03/marsmotion.html. Accessed on Mar 1, 2013.
4. http://www.cliffsnotes.com/study_guide/topicArticleId-23583,articleId-23506.html. Accessed on Apr 15, 2013.
5. Sheehan W. The planet mars: a history of observation and discovery. Tucson: The University of Arizona Press; 1996.
6. http://physicsworld.com/cws/article/news/2012/aug/17/geologist-claims-to-have-found-plate-tectonics-on-mars. Accessed on Apr 2, 2013.
7. http://www.nasa.gov/audience/foreducators/k-4/features/F_Measuring_the_Distance_Student_Pages.html. Accessed on Apr 3, 2013.
8. http://library.thinkquest.org/12145/traj.htm. Accessed on Apr 2, 2013.
9. http://www.astronomytoday.com/astronomy/mars.html. Apr 20, 2013.
10. http://quest.nasa.gov/aero/planetary/mars.html. Accessed on Apr 18, 2013.
11. http://www.ozgate.com/infobytes/mars_geology.htm. Accessed on Apr 2, 2013.
12. Mapping Mars: science, Imagination, and the birth of a world. New York: Picador USA. ISBN0-312-24551-3. p. 98.
13. http://mars.jpl.nasa.gov/programmissions/science/goal3/. Accessed Apr 12, 2013.

14. Nimmo F, Tanaka K. Early crustal evolution of mars. Ann Rev Earth Planet Sci 2005; 33:133–161, available at http://adsabs.harvard.edu/abs/2005AREPS.33.133N. Accessed on Mar 1, 2013.

15. http://science.yourdictionary.com/articles/how-many-moons-does-mars-have.html. Accessed on Mar 26, 2013.

16. http://www.nasa.gov/exploration/whyweexplore/Why_We_27.html. Accessed on Mar 22, 2013.

17. http://science.yourdictionary.com/articles/how-many-moons-does-mars-have.html. Accessed on Apr 14, 2013.

18. http://www.ifa.hawaii.edu/ ~ barnes/ASTR110L_F03/marsmotion.html. Accessed on Apr 2, 2013.

19. http://starryskies.com/The_sky/events/mars/opposition01.html. Accessed on Apr 19, 2013.

20. Aphelion/Perihelion is an object's orbital point (in distance and time) around a star where the object's distance (on its elliptical orbit) from its parent star is farthest/closest. The terms apogee & perigee are used instead when referring to objects orbiting the Earth (e.g., the Moon); a poapsis & periapsis refer stoor bits around all other bodies. http://www.astronomytoday.com/astronomy/basics.html. Accessed on Mar 24, 2013.

21. Meeus J. When was mars last this close? Planetarian 2003; pp. 12–13.

22. http://www.universetoday.com/14824/distance-from-earth-to-mars/#ixzz2QzpUWsjZ. Accessed on Apr 2, 2013.

23. The subsequent discussion is mainly based on 1. http://www.seasky.org/solarsystem/assets/animations/solar_system_menu.jpg&imgrefurl and http://www.seasky.org/. Accessed on Mar 14, 2013.

24. http://my.tbd.com/blogs/weather/archive/?year=2010&month=08. Accessed on Jun 21, 2013.

25. http://www.enchantedlearning.com/subjects/astronomy/planets/. Accessed on Jun 21, 2013.

26. http://starryskies.com/The_sky/events/mars/opposition01.html. Accessed on Jun 21, 2013.

27. http://cseligman.com/text/planets/marsoppositions.htm. Accessed on Jun 21, 2013.

28. www.wordpress.com and http://mydarksky.org. Accessed on Jun 21, 2013.

Part II
India and Mars Agenda

Chapter 4
Indian Multidimensional Space Plan

> *There are some who question the relevance of space activities in a developing nation. To us, there is no ambiguity of purpose. We do not have the fantasy of competing with the economically advanced nations in the exploration of the moon or the planets or manned space-flight.*
>
> *But we are convinced that if we are to play a meaningful role nationally, and in the community of nations, we must be second to none in the application of advanced technologies to the real problems of man and society.*
>
> Dr. Vikram Ambalal Sarabhai

India made a nascent beginning in the space arena in 1963 and today is been globally reorganised as one of the leading space powers in the world. In 1963, India's entry into the space field began with the launching of sounding rockets. India began its space programme from a small church in the southern parts of the country[1] [1]. During last five decades, India has made a rapid progress in space arena and is now in a position to undertake missions in the deep-space arena. India has successfully completed its first moon mission in 2009. Presently, India has one of the biggest networks of remote sensing satellites in the world and also has various communication satellite systems on its inventory.

During initial phase of its space programme, India took slow but steady steps. Initially, India's space programme started under the aegis of Department of Atomic Energy in 1962 with creation of Indian National Committee for Space Research (INCOSPAR). The mandate to the committee was to oversee all aspects of space research in the country [2]. The first sounding rocket by India was launched in 1963 with the help from National Aeronautics and Space Administration (NASA). India's former Prime Minister Indira Gandhi dedicated Thumba Equatorial Rocket Launching Station (TERLS) to the United Nations on Feb 2, 1968. On that occasion, INCOSPAR Chairman Dr. Vikram Sarabhai articulated India's goal in space programme. He stated that India's space program is civilian, with focus on the application of space technology as a tool for socio-economic

---

*He is known as the father of the Indian space programme and was the former Chairman ISRO, http://www.isro.org/Ourchairman/former/vsbhai.aspx, accessed on Apr 24, 2013.

---

[1] This church the only building available at Thumba village (Kerala state in southern India) was chosen because the geomagnetic equator passes through that location.

A. Lele, *Mission Mars*, SpringerBriefs in Applied Sciences and Technology, DOI: 10.1007/978-81-322-1521-9_4, © The Author(s) 2014

development of the country. The basic aim of India's space programme was described as a programme capable of using space technologies in the vital areas of development such as communications, meteorology and natural resource management [3].

Later, Indian Space Research Organisation (ISRO) was formed under the Department of Atomic Energy in 1969 and was subsequently brought under the Department of Space in 1972. A Space Commission was also set up in the same year which reports directly to the Prime Minister. In the year 1975, India conducted its first nuclear test. This made India to go under the global sanctions, and the international assistance in regard to technological assistance stopped. This resulted in seriously impacting India's progress in space. However, India converted this technological apartheid into an opportunity and began the process of indigenisation. Naturally, initially ISRO took some time to establish itself, but eventually, now India's space programme is standing on a sound footing.

India placed it's first satellite, Aryabhatta, in orbit with the help of the USSR during April 1975. India became the space-faring nation during July 1980, with the launching of a satellite using its own rocket launching system and from own soil. Since then, India has come a long way and has achieved significant successes in various fields of space technologies. India's most successful launcher the Polar orbiting Satellite Launch Vehicle (PSLV) has played a crucial role in ISRO's progress. PSLV commenced its operational launches in 1997 and since then has gained an image of most dependable workhorse with it's last 22 successive successful space launches. Today, India is in a position to launch a 2,000–2,500 kg weight satellite into geostationary orbit[2] [4]. PSLV was also instrumental in assisting India to launch its first mission to the Moon. With the help of this vehicle, ISRO in 2009 was also able to undertake multiple satellite launches in a single rocket launch. ISRO has launched 10 satellites in one go by using PSLV-C9 launcher, creating a unique record globally.

However, the major limitation of India's space programme is its inability to launch heavy satellites (say of the variety of 4,000 kg or more) to the geostationary orbit. India's Geosynchronous Satellite Launch Vehicle (GSLV) programme is yet to become operational. It had suffered two successive failures in April and December 2010. Now, India is expected to resume flying the GSLV rocket in the second half of 2013. India's basic limitation in making GSLV work is the absence of cryogenic engine technology. During 1992, Russia was stopped from parting this technology by the United States because of the fear that India could use this technology in its missile programme. Even today, India is yet to succeed in developing the cryogenic engine technology required for such launches. However, it is expected that by 2013, India would be testing this technology and expected to become self-sufficient in the area of launching satellites weighing around 4–5 tonnes.

---

[2] On Sep 2, 2007, India successfully launched its 2,500 kg INSAT-4CR geostationary satellite with GSLV F04 vehicle. For various details on India's development in space sector, refer [4].

In 1992, the ISRO has established its commercial outlet called the Antrix Corporation. This organisation markets space and telecommunication products of ISRO. India has also assisted the launch of 35 foreign satellites [5] by offering their launch facilities either on commercial terms or otherwise. Antrix Corporation provides end-to-end solution for space products, for varied applications covering communications, earth observation, and scientific missions; services including remote sensing data series, transponders' lease service; launch services; mission support services; and a host of consultancy and training services [6].

Remote sensing has been one of the areas of core competence for ISRO. It has designed and launched satellites with 80-cm resolution[3]. At global level, ISRO systems offer one of the best resolutions in this field. The communication sector is another sector where India's has significant interests. The Department of Space (DoS) has projected demand for 794 transponders in the 12th plan (2012–2017) from an operational transponder capacity of 187 [7]. Much needs to be achieved in this field, and ISRO needs to put in lot of efforts to achieve its target. India is also planning for a regional navigation system, mainly to cater for its own needs. The Indian Regional Navigational Satellite System (IRNSS) is a constellation of seven satellites and will cover only Indian and adjoining regions. This is expected to become operational by 2015, and the first satellite is expected to be launched in 2013. India is also collaborating with other states and undertaking joint missions. Particularity, India's collaboration with France is noteworthy. Both the states are found jointly developing satellites, and two such satellites have been launched, one in 2011 (Megha-Tropiques) and other in 2013 (SARAL), respectively. These systems are meant for collecting various types of meteorological observations and for studying the circulation of ocean currents.

Apart from remote sensing, meteorology, communication and navigation satellites, India has also interests in various other areas of space technologies and space science. In 2007, ISRO had also successfully launched and recovered its Space Capsule Recovery Experiment (SRE), demonstrating its thermal protection system abilities which would eventually help towards developing its human space programme. However, India has no plans to undertake the human mission in space at least in near future. In the year 2013, ISRO is proposing to launch its first dedicated astronomy satellite called Astrosat.

In the recent past, a major transformation has taken place in ISRO's space agenda. This was in the aftermath of a 2006 meeting (held at Bangalore) of top Indian scientists to discuss the future of India's space programme. The seeds of India's deep-space mission could be said to have been sown in that meeting. However, it is important to note that such high profile missions have been undertaken only as additional missions, keeping the core programme intact. ISRO has done India proud with the spectacular success of its Moon mission (2008–2009), particularly by playing a key role in finding water on the surface of the moon. India's second Moon mission is expected to be space bound in 2014.

---

[3] Cartosat-2 launched in 2007 can produce images of up to **80 cm** in **resolution.**

With the success of historic Indo-US nuclear deal in 2005, now international restrictions put on ISRO in regard to technology transfer have been removed. Particularly, the ISRO was under the United States entity list and four of its major subsidiaries were excluded from doing any technology collaborations with other international agencies [8]. These restrictions got removed in 2011. Now it is expected that they could interact with other states more freely and technology transfer and technology collaborations would start. India's investment towards utilising space for strategic purposes has been cautious. Dual-use nature of satellite technologies allows India to gain some inputs for its security establishments. However, India is strictly against the weaponisation of the outer space. It has been reported that in 2013, India would be launching a dedicated satellite for Indian Navy which is expected to be a communication satellite.

India has thus far been a very active participant in many of the international multilateral forum such as the Inter-Agency Space Debris Coordination Committee, the United Nations Committee on the Peaceful Uses of Outer Space, and the Conference on Disarmament in Geneva. India is a late entrant into the space security debate and presently appears to be in a wait-and-watch mode [9].

For a country with a long experience, significant successes, well-established infrastructure and trained scientists in space sector, it is but obvious that it would attempt more challenging and scientifically important missions. As mentioned earlier during January 2007, India tested and validated its re-entry technology with a 12 day mission of the Space Capsule Recovery Experiment (SRE-1). Subsequently, during 2008–09, India had conducted the Moon mission. The mission to Mars appears to be the next logical step.

References

1. Das SK. Touching lives. New Delhi: Penguin Books; 2007. p. 1.
2. http://www.bharat-rakshak.com/SPACE/space-history2.html. Accessed on Dec 1, 2008.
3. Sankar U. The economics of India's space programme. New Delhi: Oxford University Press; 2007. p. 1–2.
4. Lele A. Asian space race: rhetoric or reality?. Heidelberg: Springer; 2013. p. 59–67. India's Space Programme.
5. http://www.isro.org/pdf/foreignsatellite.pdf. Accessed on Apr 15, 2013.
6. http://www.antrix.gov.in/aboutus.html. Accessed on Apr 8, 2013.
7. Outcome Budget of the Department Of Space: Government of India (2012–2013), p. 16. *The Indian Express*, Oct 6, 2012.
8. http://www.isro.org/parliament/2011/Budget/LUSQ2227.pdf. Accessed on Apr 8, 2013.
9. Bharath G. Space security: India, in Crux of Asia. In: Tellis A, Mirski S, editors. Carnegie Endowment for International Peace. 2013.

Chapter 5
Mars Orbiter Mission

Once you show an affordable scale of the activity, then you qualify yourself to be a partner of international programme. So, when future manned missions or even future important missions to Mars take pace, India would be part of the global community because you have already demonstrated that you have reached the place (Mars).

Dr K. Kasturirangan

5.1 Mission Silhouette

5.1.1 India's Mars Agenda

In ancient times, the principal Indian languages then referred the planet Mars primarily as Mangala and also with names like Angaraka and Kuja. These names mean auspicious, burning coal (red in colour) and the fair one respectively. There are various references in different periods of history about Mars but mostly all lead to indicate that the Mars is a god of war. India has been a land famous for studies in astrology for many centuries [1]. During 6th Century BC, Aryabhata the famous Indian astronomer conceptualised various ideas about the solar systems and presented various mathematical formulations to support his arguments. Incidentally, India's first satellite has been named after him.

Astronomy in Asia has continuously developed. Local wisdom in many Asian countries reflects their interest in astronomy since the historical period. However, the astronomical development in each country in yesteryears had a different flavour depending on their cultures, politics and economics. Astronomical studies in countries such as China andIndia are found well developed for many centuries [2]. Since ancient times, there has been a continued interest in astronomy in India. Presently, there are few specialised agencies in India which undertake studies in various aspects of astronomy and astrophysics [3]. Also, there are societies like the Astronomical Society of India [4] (established in 1972) a prime association of professional astronomers in India. The society has close to 1,000 members. The objectives of the society are the promotion of Astronomy and related branches of science in India. The society organises scientific meetings, publishes a quarterly

---

*He is former Chairman ISRO. http://www.thehindubusinessline.com/news/science/mars-mission-to-boost-indias-global-credentials-kasturirangan/article4332033.ece. Accessed Apr 29, 2013

A. Lele, *Mission Mars*, SpringerBriefs in Applied Sciences and Technology,
DOI: 10.1007/978-81-322-1521-9_5, © The Author(s) 2014

bulletin, and supports the popularisation of Astronomy and various other relevant activities. The foremost research agency in space sciences in India is Physical Research Laboratory (PRL) which carries out fundamental research in select areas of Physics, Space and Atmospheric Sciences, Astronomy, Astrophysics and Solar Physics, and Planetary and Geosciences [5]. Under the department of space in 2007, Indian Institute of Space Science and Technology (IIST) has been formed. This is Asia's first Space Institute and the first in the world to offer the complete range of undergraduate, postgraduate, doctoral programmes with specific focus to space science, technology and applications. This institute integrates the research and development with academics and encourages faculty members to undertake research [6].

With such a well-established space science base in the country, it is but obvious that ISRO would be keen to make inroads in the area of deep space. The mission to Mars could be seen as a step in the continuation of the overall astronomy and space science agenda within the country.

Mars has been the subject of interest for various space agencies and space scientists for many years. Scientists in ISRO have been studying Mars for some time, probably initially out of interest in astronomy and because of curiosity to know more about the medium in which the rocket scientists operate. During 1960s and 1970s, the erstwhile USSR and the United States were undertaking various missions to Mars but unfortunately with low success rates. At the same time, the success of Apollo mission was a great motivator for the scientists across the world. All such happenings kept in Indian scientific community conscious of the activities in the deep space arena.

It could be difficult to exactly identify that when ISRO actually started working towards undertaking the Mars mission. However, it could be inferred that as an aftermath of the 2006 Bangalore meeting of the scientists this idea could have got germinated. Planetary science has been an important element of the ISRO's space agenda for many years. References in India media are found since 2007 about ISRO's interest in Mars mission. The ISRO Chief Mr G Madhavan Nair during Apr 2007 had made a statement to the effect that if India starts the preparations for the Mars mission immediately then within 5 years they could succeed in their intend. A year later, he has said that "The study for Mars exploration had already started. It is expected to take at least 4 years to complete the initial studies to take up the Mars Mission" [7]. During 2006, the previous Chief of the ISRO Mr. K Kasturirangan after the signing of the Indo-US space deal had mentioned that ISRO and NASA for future joint space explorations could include Mars and other inter-planetary missions [8].

Hence, although the official announcement of the India's November 2013 mission was made on Aug 15, 2012, the preparations could be said to have began much earlier. In fact during September 2012, the Hindustan Aeronautics Limited (HAL) has handed over the Mars Orbiter Mission satellite structure to the ISRO's Satellite Centre. This satellite structure is an assembly of composite and metallic honeycomb sandwich panels with a central composite cylinder. ISRO is building the other satellite subsystems and scientific payload onto this structure [9]. Overall, ISRO is found concentrating on its Mars programme in a planed manner.

It is extremely important to appreciate that reaching Mars has a larger strategic significance. Former President of India Dr. APJ Abdul Kalam has articulated the importance of India's Mars project very succinctly. He emphasises that, "Mars is international property; all the planets belong to the international community. It is essential to establish that we have done our job and our job has important scientific goals and we should do that only then we can say then that Mars belongs to us" [10].

5.1.2 Mission Mars

For ISRO, November 2013 would be its first attempt to reach Mars. It is important to note that India has so far undertaken only one deep space mission, its mission to Moon in 2008. However, the challenges for Mars are much more complicated than the Moon. The challenges are wide-ranging from distance to gravity to radiation. Communication systems are among the most critical functions for any form of space exploration and more so for the deep space missions. For such missions, the enormous distances up to tens of billions of kilometres from the earth pose major challenge. Also, in the present era, because of the technological advancements in the space based sensors sector there is an increasing demand for the transmission of the different types and forms of data simultaneously and that too almost in real time. In deep space arena, real-time data transfer is not possible however; for the present generation sensors in the deep space arena (say Mars), the return data rates (number of bits per second) have increased significantly and all this requires state-of-art communication infrastructure.

Straightforwardly, India's mission involves sending a satellite in the close vicinity of Mars where it is expected to function as an observation platform. Normally, every remote sensing satellite performs similar functions, but the only difference is that it observes earth from space and in this case it is going to observe the Mars. There are certain medium related problems too. The rocket-carrying remote sensing satellites delivers them in low earth orbit to their preplanned destination in say less than half hour's time after the rocket is launched from the launch site and subsequently such satellites are declared operational within few days. In case of Mars, the present level of technology demands approximately 300 days of time after the launch from the earth to reach the destination. Subsequently, the orbit insertion manoeuvres are conducted.

ISRO's proposed Mars mission involves various intricate stages and each stage has its own set of challenges. In general, the mission could be subdivided as launch phase, travel phase (300 days approximately), orbit insertion phase and observation phase. Communications would be the key for all these phases. The mission has its own priorities, and there is clarity in regards to both the technological and scientific aspects of the mission. The mission is basically about developing the technology to reach Mars and seeking answers to fundamental questions like origin and the evolution of the red planet.

As of July 2013 being in the final stages of preparation for the mission, the ISRO is involved in undertaking various activates. The planning aspects of the mission are already over and now the phase of execution is under progress. Following are some of the important areas where the organisation has paid/is paying special attention for the success of this mission:

- Design, development and manufacture a Mars Orbiter (an Indian public sector organisation Hindustan Aeronautics Limited has played an important role in this development). Identification of the scientific objectives of the mission and accordingly design and develop the specific sensors for the mission for further integration with the orbiter.
- Arrange the communication facilities by developing a deep space network. Identify the locations in the various parts of the world where the communication silos would be required to be organised.
- Identify the suitable launch vehicle and prepare the rocket for the launch of this mission.
- Post successful mission launch plan for monitoring the health of the mission during various further phases of the mission. Undertake various orbiting raising manoeuvres as per the plan.
- Plan for any potential midcourse corrections.
- Plan for the data collection after the Mars Orbiter becomes fully operational and continuously monitor the process of data collection.
- Develop a mechanism for timely public dissemination of information about the various mission related activities.

5.1.3 The Mars Orbiter Mission

Mars Orbiter Mission is ISRO's first interplanetary mission to planet Mars with an orbiter craft designed to orbit Mars in an elliptical orbit. The mission is primarily a technological mission considering the critical mission operations and stringent requirements on propulsion and other bus systems of the spacecraft. It has been configured to carry out observation of physical features of Mars and carry out limited study of Martian atmosphere with five payloads onboard.

5.1.3.1 Mission Objectives

The main objectives are to develop the technologies required for design, planning, management and operations of an interplanetary mission comprising the following major tasks:

- Orbit manoeuvres to transfer the spacecraft from Earth-centred orbit to helio-centric trajectory and finally capture into Martian orbit

- Development of force models and algorithms for orbit and attitude computations and analyses
- Navigation in all phases
- Maintain the spacecraft in all phases of the mission meeting power
- Communications, thermal and payload operation requirements
- Incorporate autonomous features to handle contingency situations.

5.1.3.2 Scientific Objectives

The scientific objectives deal with the following major aspects:

- Exploration of Mars surface features by studying the morphology, topography and mineralogy using specific scientific instruments.
- Study the constituents of Martian atmosphere like methane, CO_2, etc. using remote sensing techniques
- Study the dynamics of upper atmosphere of Mars, effects of solar winds and radiation and the escape of volatiles to space.

The mission would also provide multiple opportunities to observe Mars moon "Phobos" and also offers an opportunity to identify and reestimate the orbits of asteroids seen during the Martian Transfer Trajectory (MTT).

5.1.4 Launch Platform

India proposes to use its Polar Satellite Launch Vehicle (PSLV) [11] for the launch of mission Mars. PSLV is the most successful operational launch vehicle of ISRO. PSLV has successfully undertaken Sun-synchronous, Geo-synchronous Transfer Orbit and low inclination missions in the past. As off Jun 2013, PSLV has launched 62 satellites/spacecraft (27 Indian and 35 Foreign Satellites) into a variety of orbits so far. There had been 22 continuously successful flights of PSLV, till February 2013. Approximately a single PSLV launch costs 17 million USD. PSLV is capable of launching 1,600 kg satellites in 620 km sun-synchronous polar orbit and 1,050 kg satellite in geo-synchronous transfer orbit. In the standard configuration, it measures 44.4 m tall, with a lift-off weight of 295 tonnes. PSLV has four stages using solid and liquid propulsion systems alternately. To undertake the PSLV launch in standard mode, a cluster of six strap-ons is attached to the first stage motor, four of which are ignited on the ground and two are air-lit. The core-alone (CA) model does not include the six strap-on boosters used by the PSLV standard variant. This model allows the vehicle to carry relatively lighter satellites. On the other hand, PSLV-XL version is boosted by more powerful, stretched strap-on boosters and allows putting additional 200 kg payload (total 1,800 kg) in space. Weighing 320 tonnes at lift-off, this vehicle carries 12 tonnes of solid propellants against 9 tonnes in standard configuration. The first version of PSLV-XL had successfully launched India's first Moon mission called Chandrayaan-1.

The success of Chandrayaan-1 has demonstrated that ISRO's existing launch vehicle technology with some additional modifications is capable of undertaking an unmanned mission to a celestial body. ISRO has the capability of launching spacecraft to Mars with the existing Polar Satellite Launch Vehicle (PSLV-XL), which is a proven launcher technology. To launch a spacecraft to Mars, in general, two major options could be considered: one, a fly-by mission, and second, the orbiter missions. Comparatively, the complicities are more with the orbiter missions. However, an orbiter, in an orbit of very long time period, offers the opportunity to study Mars for a considerably longer period as compared to a spacecraft in a fly-by mission. Hence, ISRO went in the favour of planning Mars orbiter capability. It needs to be noted that the PSLV has few limitations to undertake missions like Mars mission particularly it has very limited weight carrying capability. However, probably because of the GSLV not being the time-tested vehicle and with GSLV-MkIII being still under development ISRO had very limited options. Hence, they have designed a suitable trajectory such that PSLV-XL can achieve the objective with a modest payload capability.

Mission planers had to study several possible scenarios to take the Mars orbiter to reach Mars. Options like circular Low Earth Orbit (LEO) and Elliptic Parking Orbit (EPO) were considered as parking orbits and the merits and demerits were studied. It was found that parking a spacecraft in an elliptic orbit has certain distinct advantages. Different methods of reaching Mars, like stage separation in LEO before departure, method of using the same Liquid Engine Motor (LEM) for departure, trajectory correction and arrival, staging of spacecraft and its separation in Mars Transfer Trajectory (MTT) after imparting required departure velocity in the EPO approach etc., were considered. The launch opportunities and the corresponding velocity requirements for direct ballistic trajectories are also considered. For these opportunities, payload capabilities are assessed for orbiter missions to planet Mars.

The basic requirement in regards to mission design and adopting manoeuvring strategy is to achieve the desired trans-Mars trajectory with minimal fuel composition. This could be achieved by undertaking direct or indirect transfer methods. The direct transfer can be further classified into traditional and unconventional methods. A conventional direct transfer to Mars essentially puts the spacecraft into a transfer trajectory to Mars in one go from a parking orbit by providing a large thrust and achieving the required incremental velocity in a short but finite duration. The parking orbit used for departure from Earth is generally a circular or slightly elliptic LEO. All the Mars Missions from 1960s to the 2000s used this traditional approach. This type of trajectory is used for transfer from parking orbit to heliocentric trajectory near the departure planet and from heliocentric trajectory to parking orbit near the arrival planet respectively.

An unconventional direct ballistic transfer can be from a highly elliptic parking orbit. Since the energy of high EPO is considerably higher than that of LEO, there is a reduction in the velocity requirement to inject the spacecraft into the hyperbolic trajectory. Thus, the launch vehicle itself provides substantial energy and that required from the spacecraft is correspondingly a smaller amount. The spacecraft is inserted into Mars Transfer Trajectory by a dedicated stage or by the second burn of the upper stage of the launch vehicle. In both the cases, thrust is given and required incremental velocity is achieved at the perigee of the parking orbit. However, this can be implemented either with large thrust in a short but finite duration at one stretch or with moderate thrust with multiple firings at the perigee of the successive orbits.

Indirect methods employ the following techniques individually or in combination which lower the velocity requirement but involve longer transfer time; gravity assist from other celestial bodies to reach Mars, very low continuous thrust for a considerably longer duration of the order of several months and use of weak stability boundary approach wherein the spacecraft travels to the Lagrangian [12][1] regions of Earth–Sun system. Because of lower energy requirement, the indirect methods are being increasingly considered for planetary mission design. However, for mission to Mars, trajectories with gravity assist from Venus does not offer competitive benefits in terms of reducing the velocity requirements. The trajectories with Venus–Earth Gravity assist (VEGA) and Lunar swing by offer only marginal advantage for departure. But the flight time and arrival energies of these trajectories are larger than those of ballistic trajectories. It is known that a direct ballistic transfer from highly elliptic parking orbit greatly simplifies the developmental requirements for low cost missions as demonstrated in Chandrayaan-1 mission and is the best-suited approach for unmanned planetary mission using PSLV-XL in the Indian context.

For November 2013, Mars mission launch the ISRO would be using PSLV-XL, a novel mission (C-25) design carried out, wherein a spacecraft mass of 1,350 kg would be placed in an elliptical orbit of 370 km by 80,000 km. This means that the mission would go round the Mars in an elliptical path closest at 370 km and farthest at 80,000 km. Total five different sensors with a combined payload of 15 kg would be undertaking various observations.

5.1.5 Mars Orbiter

Exploring Mars by using orbiters has its own advantages. In the past, orbiters circling Mars have provided significant information in respect of canyons, volcanoes, craters, gullies and runoff channels, clouds, weather patterns, rocks, hills, polar ice caps, eclipses, and more. They have helped in increasing the knowledge about the red planet's atmosphere, landforms, gravity, magnetic fields, elemental

---

[1] Lagrange points are locations in space where gravitational forces and the orbital motion of a body balance each other. There are five Lagrangian points in the Sun-Earth system and such points also exist in the Earth-Moon system. Refer [12]. Such points could serve as staging hubs for deep-space exploration in future.

and mineral composition, internal structure, and weather. Orbiters could have unique designs to assist even in communications and navigation. Particularly an orbiter and rover combination could assist in conducting various scientific experimentations. It becomes difficult for the rover to send the data directly to the earth station and an orbiter could play a key role as communications relays. For futuristic missions in coming decades, orbiter could even play a role for sample return missions [13]. India's orbiter could be viewed as a versatile spacecraft with a well defied role.

The spacecraft configuration for Mars mission is derived from Chandrayaan-1 heritage, which is a balanced mix of design from flight proven IRS/INSAT bus. While the solar array, TTC-RF, data transmission and thermal systems are specific to Mars mission, other aspects of system design have flight proven heritage. Modification required for Mars mission is in the areas of communication and power, propulsion system (mainly related to liquid engine restart after nearly a year) and mechanism. Following diagrams offer an idea about how various systems would be integrated (Figs. 5.1, 5.2, 5.3 and 5.4).

All the primary structure, equipment panels and special brackets for this craft are fabricated as per ISRO design in dedicated facilities established for ISRO by HAL. The 390-L-capacity propellant tanks used for Chandrayaan-1 accommodate a maximum of 850 kg of propellant which is adequate for the proposed Mars

Fig. 5.1 Stowed view

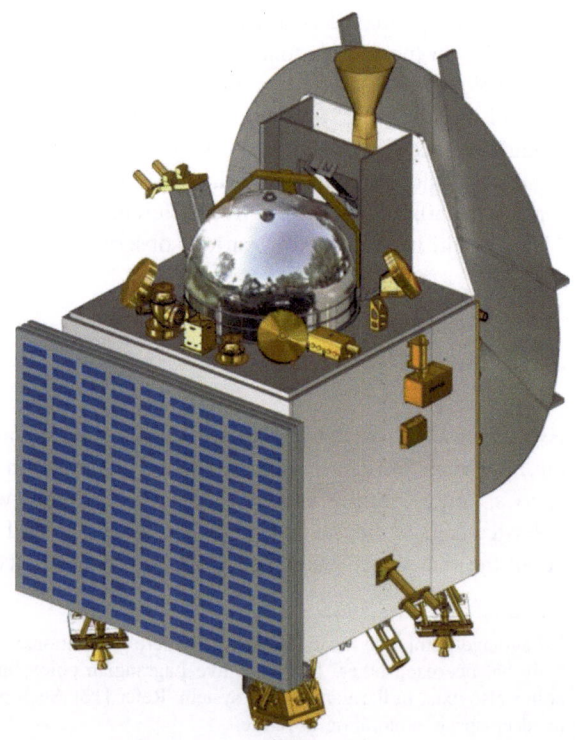

Fig. 5.2 Launch
configuration

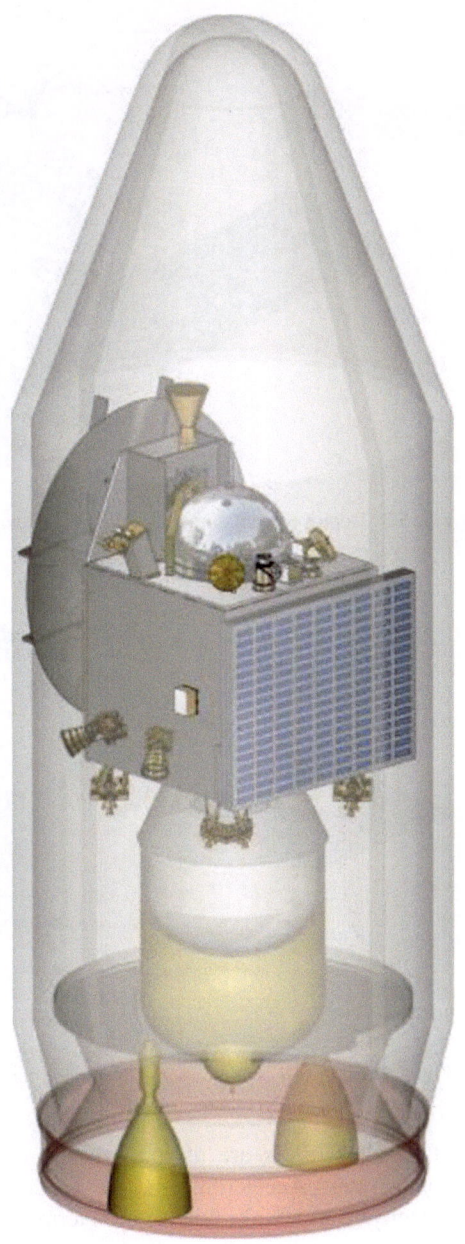

mission. A Liquid Engine of 440 N thrust is planned to be used for orbit raising
and Martian Orbit Insertion (MOI). Additional flow lines and valves have been
incorporated to ensure LE 440 N engine restart after 300 days of Martian Transfer
Trajectory (MTT) cruise and to take care of fuel migration issues. Eight numbers

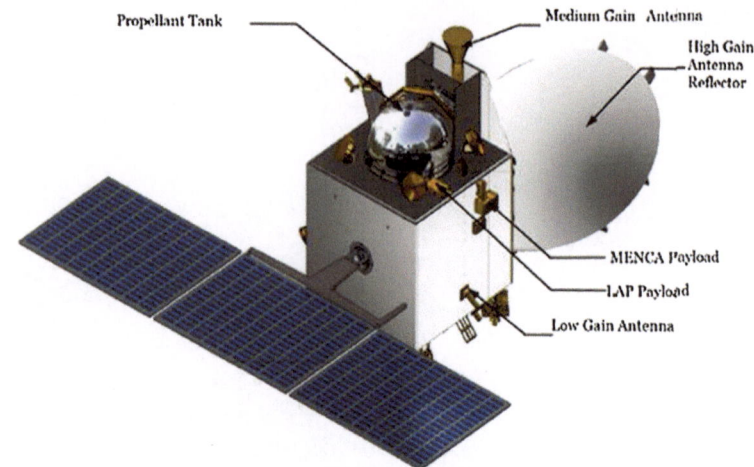

Fig. 5.3 Deployed configuration—displaying two payloads and other details

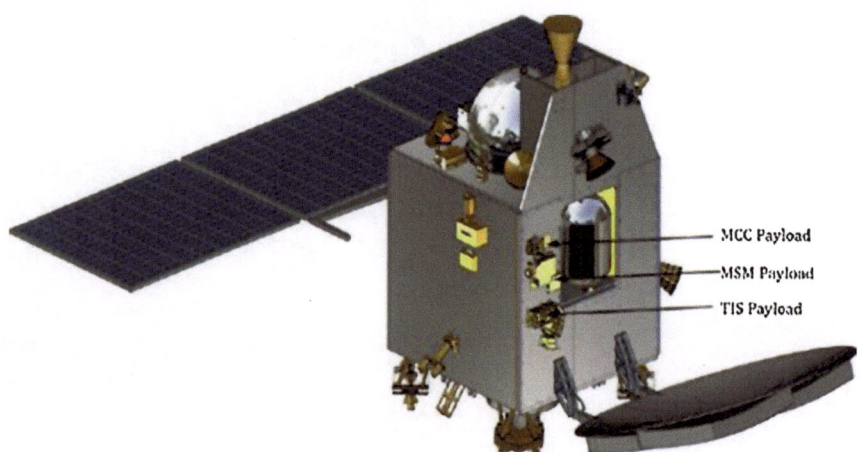

Fig. 5.4 Deployed configuration—displaying three payloads

of 22 N thrusters are used for wheel desaturation and altitude control during manoeuvres. Accelerometers are used for measuring the precise incremental velocity (ΔV) and for precise burn termination. Star sensors and gyros provide the altitude control signals in all phases of mission.

The solar panel requirement for Chandrayaan-1 was a panel of size 1,800 × 2,150 mm. However, to compensate for the lower solar irradiance (50 % compared to Earth), the Mars orbiter would require three solar panels of size 1,400 × 1,800 mm. Single 36AH Li-Ion battery (similar to Chandrayaan-1) is sufficient to take care of eclipses encountered during Earth-bound phase and in Mars orbit.

The communication dish antenna is fixed to spacecraft body. The antenna diameter is 2.2 m which is arrived after the trade-off study between antenna diameter and accommodation within the PSLV-XL envelope. On-board autonomy, functions are planned as the large Earth–Mars distance does not permit real-time interventions. This will also take care of on-board contingencies. The spacecraft bus system weighs approximately 501 kg.

5.1.6 Mission Plan

A space mission starts with an objective which could be a science objective, a technology objective, a political objective, or some combination of the three [14]. Mission planning and mission execution process is a very elaborate process and would involve very minute details above various aspects of the mission. The purpose here is not to provide any such detailed narration about India's Mars mission but to provide only a very brief idea about how the scientific community had identified and catered for various important stages of evolution.

Basically, ISRO undertook the mission planning in conjunction with the defined mission objectives. The Mars Mission is envisaged as a rendezvous problem, wherein the spacecraft would be injected into an elliptic parking orbit by the launcher. With six main engine burns, the spacecraft would be gradually manoeuvred into a departure hyperbolic trajectory to allow it to escape from the Earth's Sphere of Influence (SOI). The SOI of earth ends at 9,18,347 km from the surface of the earth beyond which the perturbing force on the orbiter is due to the sun only. Following diagram provides an indication about the 300 day travel, locations of earth and Mars during this period and the path followed by the craft towards for achieving Mars orbit insertion (Fig. 5.5).

One primary concern of this mission is to get the spacecraft to Mars, on the least amount of fuel. ISRO would be using a method of travel called a Hohmann Transfer Orbit [15][2]—or a Minimum Energy Transfer Orbit—to send a spacecraft from Earth to Mars with the least amount of fuel possible. The spacecraft leaves Earth in a direction tangential to its orbit and encounters Mars tangentially to its orbit. The flight path is roughly one half of an ellipse around sun. Eventually, it will intersect the orbit of Mars at the exact moment when the Mars is there too. This route becomes possible with certain allowances when the relative position of Earth, Mars and Sun forms an angle of approximately 44 degrees. Such an arrangement recurs

---

[2] To launch a spacecraft from Earth to an outer planet such as Mars using the least propellant possible, first consider that the spacecraft is already in solar orbit as it sits on the launch pad. This existing solar orbit must be adjusted to cause it to take the spacecraft to Mars: The desired orbit's perihelion (closest approach to the sun) will be at the distance of Earth's orbit, and the aphelion (farthest distance from the sun) will be at the distance of Mars' orbit. This is called a Hohmann Transfer orbit.

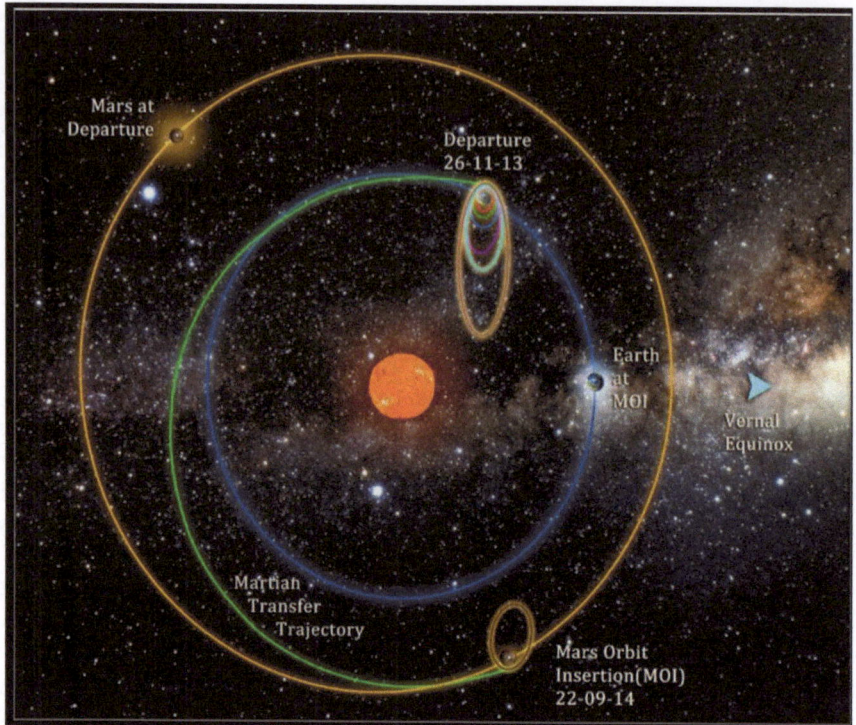

Fig. 5.5 Mission Profile

periodically at intervals of about 780 days. Minimum energy opportunities for Earth–Mars occur in November 2013, January 2016, May 2018 etc.

The spacecraft would arrive at the Mars Sphere of Influence (around 5,73,473 km from the surface of Mars) in a hyperbolic trajectory. At the time the spacecraft would reach to the closest approach to Mars (Periapsis), and it would be captured into the planned orbit around Mars with the Mars orbit insertion manoeuvre (MOI).

5.1.7 Mission Sequence

The diagram below illustrates the Mars Orbiter Mission sequence (Fig. 5.6).

With PSLV-XL as the launcher, a mission is expected to lift-off in November 2013 for a studying various facets of Mars and its atmosphere. It would be the minimum energy transfer mission, maximising the mission life providing a meaningful sustainable mission of 10 months or more, with a possibility of aerobraking at the end-of-life. Constrained to the lift-off mass of 1,350 kg and the

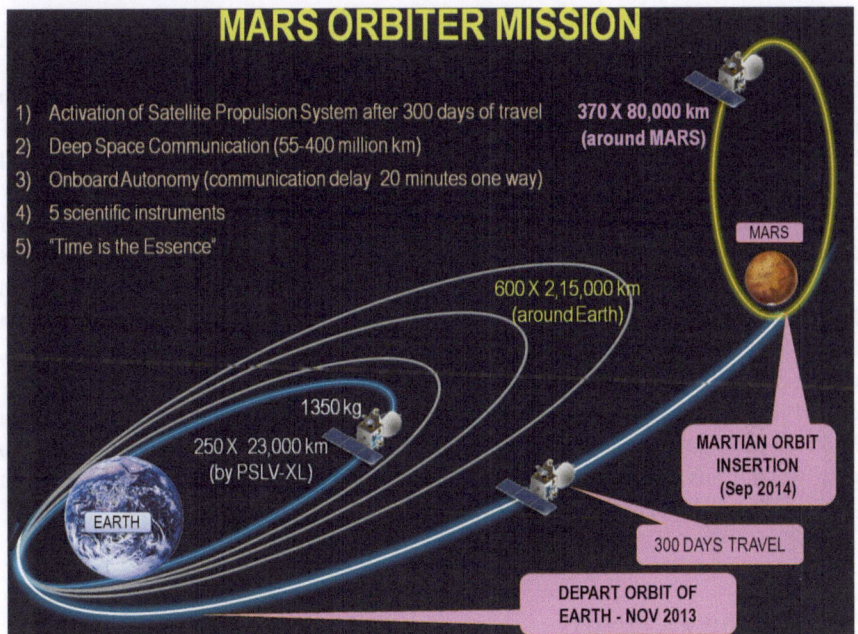

Fig. 5.6 Mars orbiter mission

propellant mass of 850 kg, a Martian orbit of $370 \times 80{,}000$ km, $32°$ inclination is achievable. The orbital period is about 3.66 days and periapsis[3] is contained in the sunlit portion of the orbit.

PSLV-XL provides Earth Parking Orbit of 250 x 23,000 km with an inclination $17°$, argument of perigee $295°$ with a spacecraft mass of 1,350 kg. The November 2013 departure conditions to Mars require an additional velocity of ~ 1.5 km/s to enter Mars Transfer Trajectory (MTT) to reach Mars. These velocity increments are imparted through a series of orbit raising manoeuvres around the perigee, raising the apogee sequentially to 40,000, 80,000 and 2,15,000 km. Subsequently, to achieve the precise departure conditions in terms of velocity and time of departure, a final perigee burn is performed placing the spacecraft in MTT.

Midcourse corrections (if any) are carried out in the MTT phase to achieve the targeted arrival conditions at Mars. Mars Orbit Insertion (MOI) is achieved by performing a deboost manuever of ~ 1.1 km/s at the periapsis of the arrival hyperbolic trajectory.

As the minimum energy transfer opportunity from Earth to Mars occurs once in 26 months, the opportunity in 2013 demands a cumulative incremental velocity of 2.592 km/sec.

---

[3] The point at which an orbiting object is closest to the body it is orbiting. As per the Kepler's laws of planetary motion, an object is at its greatest velocity at the periapsis.

5.1.8 Mission Life

The targeted mission life is 6 months after Mars Orbit Insertion. The mission can be extended if non-renewable resources in the spacecraft last for more time. Also, Spacecraft is configured with autonomous features which will ensure the survival of the spacecraft in case of a blackout or whiteout. A communication "blackout" occurs when the sun is between Earth and Mars and no voice or data link can occur for that period of time. The maximum duration of the blackout is around 17 days (Jun 6–22, 2015). A "whiteout" occurs when the Earth is between the sun and Mars and too much solar radiation may make it impossible to communicate with Earth. The maximum duration is around 14 days (May 16–29, 2016). In general, the mission is expected to function till March 2015 (would become operational by end of September 2014). All this essentially indicates that for undertaking second Mars mission during next available window in January 2016 ISRO would have to plan well in advance and base their technical designing of the mission based on some minimal inputs from its first mission.

5.2 Mission Payloads

The term payload could be defined as the carrying capacity of a spacecraft and this could include cargo, extra fuel and scientific instruments. Normally, in case of a satellite, the term payload (s) and scientific instruments onboard of that satellite is mostly found used interchangeably. These payloads perform the core functions of the satellite and are designed for wide range of applications from civil to commercial to military. In short, the payload is the cause for the satellite being there. The objective of this section is to describe and discuss the scientific instruments onboard of the Indian Mars Mission.

Over the years, the payload sensors for various satellites have been designed for specific purposes like photography, meteorological information and reconnaissance. Sensors normally undertake the measurements of various signals like light, radiation etc. which in turn help detecting, identifying and classifying objects on Earth. Various payloads have various capabilities and some of them can operate in adverse weather conductions and both during day and night.

Optimal selection and specification of satellite payloads is the biggest challenge for any organisation deciding to embark on satellite mission. Each payload is designed to perform certain functions over its useful lifetime, and mission planners must decide in advance which capabilities (i.e. which payloads) to include on new satellite launches. These decisions are complicated because [16]

1. Different payload types have different levels of importance
2. Satellite payloads age and deteriorate over time due to the harsh space environment in which they operate

3. All payload launch decisions are subject to various constraints form weight to budget
4. The selected payloads must be assigned certain engineering specifications to ensure their compatibility with the satellite bus
5. The composition of payload should match with the overall mission objective.

Mission planners usually have a very tricky job at hand particularly when they are planning a new mission. They have to determine an optimal set of payloads which would make the mission most effective. While selecting the payloads, it is important to cater for various specifications and other related requirements. They need to work minutely to decide on engineering specifications, reliability specifications, functional specifications and resource requirements.

In case of planetary missions, the decision on payload is directly dictated by the Mission objectives like in any other case. However, since we know very little abound the overall composition of the planets, it is important to include payloads which eventually would increase our knowledge about that planate.

Hence, payloads needs to be decided in such fashion that they would help to know more about the structure of the planet and nature of surface, meteorological conductions in and around the planet, various geological aspects of the planet including its mineral composition, etc. Further there are important issues like deciding on sensors required for purposes of direct and indirect mapping of the planet and study the surface of the planet further form the point of view of selecting future landings for robotic and human landings, identifying the presence (or absence) of water and presence of life, etc.

For India, there could have been multiple challenges for the selection of the payloads. This being India's first mission to Mars the real challenge is to reach to such a long distance and then undertake observations. Naturally, the focus of the mission appears to be more on the technological aspect of the travel. However, the mission also has a well-developed scientific agenda and five specifically designed payloads have been identified as the travels for this mission. ISRO had received nearly twenty scientific payload proposals from various centres of ISRO and department of space. The basic purpose behind all these payloads has been to address the science of understanding the Mars atmosphere and its dynamics.

The science objectives of the proposals focus on two major aspects: one, to look at the surface features like morphology, topography and mineralogy and second, to study the atmospheric reservoir, composition of gas, dust, ice, clouds and their dynamics. Also, the aim is to know the interaction of atmosphere with solar radiation and the resultant photochemistry, plasma interactions and loss processes. Most proposals have a mass budget within 2–3 kg.

The proposals could be broadly classified into following categories:

- Martian atmospheric studies and upper region of the atmosphere-related studies
- Solar and X-ray spectroscopy
- Mars imaging.

Based on the technical and scientific appreciation of their mission and giving due credit to the pronounced scientific aims of the mission, finally, five payloads have been selected by ISRO to visit Mars.

For the November 2013 mission, there would be three electro-optical payloads operating in the visible and thermal infrared spectral ranges, a photometer to sense the Mars atmosphere and surface and a mass spectrometer. These payloads are as follows:

1. Lyman Alpha Photometer (LAP)
2. Methane Sensor for Mars (MSM)
3. Mars Exospheric Neutral Composition Analyser (MENCA)
4. Mars Color Camera (MCC)
5. Thermal Imaging Spectrometer (TIS).

Based on the overall science objectives and considering the constraints of the orbit, above five indigenous experiments have been short-listed. The list [17] of payloads indicating purpose, mass and objectives are summarised in the Table 5.1.

Electro-optical payloads are generally used for various imaging applications. Such systems operate by modification of the optical properties of a material by an electric field. Such systems normally cater for a wide range of applications and are used for the development of various remote sensing systems. Electro-optical sensors convert light into electric single. Normally, such payloads provide stable, high resolution day and night observation capabilities but have certain limitations in respect of weather. They are found being used for different purposes like forest fires monitoring, for electronic flash unit in photography to military applications like fitted on platforms like unmanned aerial vehicles (UAVs), used for the various intelligence, surveillance, and reconnaissance and also for target identification and tracking purposes. Payloads like visible and infrared thermal imaging spectrometers are the hyperspectral imaging spectrometers that operate in the visible and mid-infrared regions. Such systems are used for remote sensing in deep space missions. Interestingly, ground-based electro-optical deep space surveillance systems are also available.

ISRO has an independent Laboratory for Electro Optics Systems (LEOS) [18]. This unit of ISRO is engaged in design, development and production of Electro-Optic sensors and camera optics for satellites and launch vehicles. The sensors include star trackers, earth sensors, sun sensors and processing electronics. Optics Systems include both reflective and refractive optics for remote sensing and meteorological payloads. Other optical elements developed by LEOS for in-house use include optics for star sensor, optics for Lunar Laser Ranging Instrument (LLRI), optical masks for sun sensors, optical filters and encoders. One of the main focuses of this organisation is to undertake research and development in new technologies for present/future satellites.

The Space Application Centre (SAC) of ISRO has provided the majority of the payloads for this mission. SAC focuses on the design of space-borne instruments for ISRO missions and development and operationalisation of applications of space technology for national development. The applications cover communication,

Table 5.1 Payload summary

Science theme	Payload	Primary objective	Mass (kg)	Development centre
Atmospheric studies	Lyman alpha photometer (LAP)	It measures the relative abundance of deuterium and hydrogen Measure deuterium/hydrogen (D/H) ratio	1.5	Laboratory of electro-optical systems (LEOS)—Bangalore
	Methane sensor for mars (MSM)	Measures methane (CH4) in the martian atmosphere with high level of accuracy	3.0	Space applications centre (SAC)—Ahmedabad
Plasma and particle environment studies	Mars exospheric neutral composition analyser (MENCA)	Map neutral composition in exosphere, martian upper atmosphere	4.3	Vikram Sarabhai space centre (VSSC)—Trivandrum
Surface Imaging studies	Mars color camera (MCC)	Optical colour imaging. It will take pictures in red, green and blue colours. The camera will help to understand Martian dust storms or dust devils	1.4	Space applications centre (SAC)—Ahmedabad
	TIR imaging spectrometer (TIR)	Thermal remote sensing. It will map the surface and mineral composition of Mars	4	Space applications centre (SAC)—Ahmedabad

broadcasting, navigation, disaster monitoring, meteorology, oceanography, environment monitoring and natural resources survey. SAC designs and develops all the transponders for the INSAT and GSAT series of communication satellites and the optical and microwave sensors for IRS series of remote sensing satellites [19].

Vikram Sarabhai Space Centre (VSSC) is responsible for one payload. In fact, this centre of ISRO is its lead Centre for launch vehicles, with its expertise in design, development and realisation of sounding rockets and launch vehicles. Along with this primary responsibility, they are also involved in developing various space application programmes [20]. On the whole, various units of ISRO are found actively involved in progressing India's Mars dream.

5.2.1 Lyman Alpha Photometer

Lyman Alpha Photometer (LAP) is an absorption cell photometer. It measures the relative abundance of deuterium[4] and hydrogen from Lyman-alpha emission in the Martian upper atmosphere (typically Exosphere and exobase). Measurement of D/H (Deuterium to Hydrogen abundance Ratio) allows us to understand especially the loss process of water from the planet (Fig. 5.7).

The objectives of this instrument are as follows:

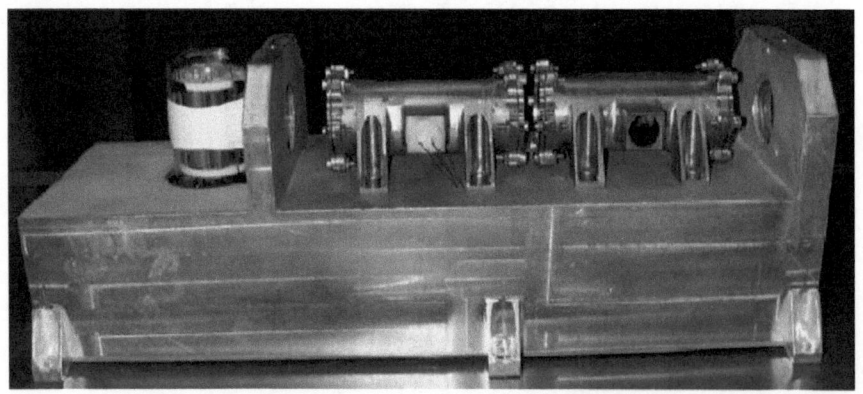

Fig. 5.7 View of LAP

---

[4] Deuterium is an isotope of hydrogen and it is believed by many that nearly all deuterium found in nature was produced in the Big Bang. There is an opinion that in future when human colonisation of Mars happens, this planet may enjoy a power-rich economy based upon exploitation of its large domestic resources of deuterium fuel for fusion reactors. Deuterium is expected to be in abundance over Mars.

(a) Estimation of D/H ratio
(b) Estimation of escape flux of H2 corona
(c) Generation of Hydrogen and Deuterium coronal profiles.

Nominal plan to operate LAP is between the ranges of approximately 3,000 km before Mars periapsis to 3,000 km after Mars periapsis. Minimum observation duration for achieving LAP's science goals is 60 min per orbit during normal range of operation.

5.2.2 Methane Sensor for Mars

This payload is a Methane sensor. MSM is designed to measure Methane (CH4) in the Martian atmosphere with ppb (parts-per-billion) accuracy and map its sources. Data is acquired only over illuminated scene as the sensor measures reflected solar radiation. Methane concentration in the Martian atmosphere undergoes spatial and temporal variations. Hence, global data are collected during every orbit. Since Field of View (FOV) of MSM is less, scanning is essential. Seven Apoareion Imaging scans of entire disc and Periareion Imaging are planned as it scans over the Periareion in every orbit (Fig. 5.8).

Presence of trace amounts of methane (CH_4) on Mars has been reported by some of the NASA missions during the last decade. Currently, the presence of Curiosity rover is assisting to know more about the methane aspects of Mars. The reasons for the origin of Mars methane could be non-biological or biological processes. There are three possible sources of Methane on Mars; cometary impacts, geology and biology. Scientists are working to indentify for the presence of methane. If the reasons are biological, then further research to know more about the presence of life on Mars (or had it existed in the planet's past) would have to be undertaken.

Fig. 5.8 MSM view

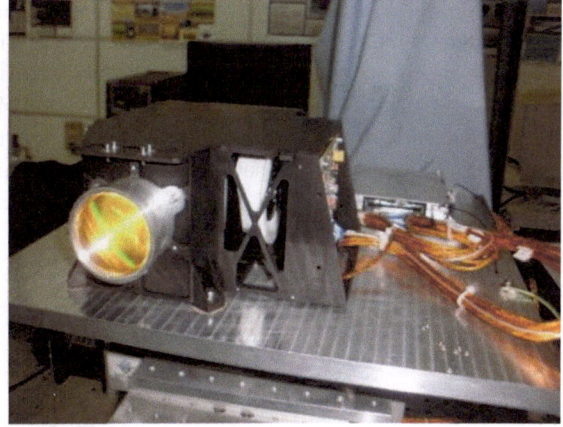

5.2.3 Mars Exospheric Neutral Composition Analyser

MENCA is a quadruple mass spectrometer capable of analysing the neutral
composition in the range of 1–300 amu (atomic mass unit) with unit mass resolution. The heritage of this payload is from Chandra's Altitudinal Composition
Explorer (CHANCE) payload aboard the Moon Impact Probe (MIP) in Chandra-
yan-1 mission. MENCA is planned to perform five observations per orbit, 1 h/
observation (Fig. 5.9).

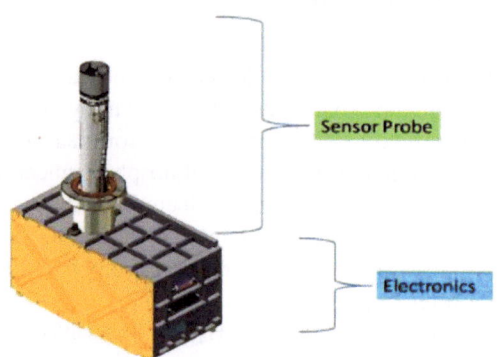

The schematic of the MENCA payload (without the DC/DC converter and the
mounting accessories) .

Fig. 5.9 Details of MENCA

5.2.4 Mars Colour Camera

This tri-colour Mars Colour camera gives images and information about the surface features and composition of Martian surface. They are useful to monitor the
dynamic events and weather of Mars like dust storms/atmospheric turbidity. MCC
will also be used for probing the two satellites of Mars—Phobos and Deimos. It
also provides the context information for other science payloads. MCC would be
providing the context information for other science payloads. MCC images are to
be acquired whenever MSM and TIS data is acquired. Seven Apoareion Imaging
of entire disc and multiple Periareion images of 540 × 540 km snaps are planned
in every orbit (Fig. 5.10).

Fig. 5.10 MCC equipment

5.2.5 Thermal Infrared Imaging Spectrometer (TIS)-Backup Payload

TIS measures the thermal emission and can be operated during both day and night. It would map surface composition and mineralogy of Mars and also monitor atmospheric CO_2 and turbidity (required for the correction of MSM data). Temperature and emissivity are the two basic physical parameters estimated from thermal emission measurement. Many minerals and soil types have characteristic spectra in TIR region. TIS can map surface composition and mineralogy of Mars (Fig. 5.11).

Fig. 5.11 TIS assembly

5.2.6 *Processing and Distribution of Data*

It is important to have a primary operating level interface to collect, process and distribute the data to the end user and also develop an archiving system. ISRO has planned its mission in such a way that imaging data of every orbit can be down linked during a single orbital period of 76 h around Mars.

ISRO had established the Indian Space Science Data Centre (ISSDC) the infrastructure established during Chandrayaan-1 and is earmarked for Mars Science data archival and accruals. ISSDC facilitates science data processing, archival, and dissemination functions for Scientists. The data transfer system at ISSDC, with suitable security systems, provides for distributing science data (as per data policy) to the concerned. The communications infrastructure is elaborate and caters to the needs of Principal Investigators (PIs) and Payload Operations Centres (POCs) [21][5]. ISSDC also interfaces with POCs for routing the operations needs of the instrument to the Spacecraft Control Centre (SCC). Level-0 and Level-1 data products of instruments, as applicable, could be routinely produced at ISSDC. Automation in the entire chain of data processing is planned. Provision is made to host higher-level data products for any instrument as supplied by the PI teams. ISSDC also provides for data archives in Planetary Data Systems (PDS) format, an international standard. The data dissemination will also follow PDS standard. The computer networking at ISSDC caters to connectivity to the IDSN operations facility, SCC and POCs. The storage and server capacities are built at ISSDC to cater to the needs of Chandrayaan-1 and other science missions to follow and for Mars mission some modifications are already been put in place[6]. Scalability and high-level availability are given high priority in architectural design. It is also planned to have a help-desk facility at ISSDC to cater to the data needs of National and International Scientists. ISSDC will also host a Website for collecting the observations needs of the scientists and making them available for an apex committee to review for further clearance for operations, etc. ISSDC data archival and distribution functions shall follow the data policy guidelines of ISRO.

The ISSDC will be the single nodal agency to distribute scientific data to all the user scientists nationally and internationally on demand.

The main responsibility of the ISSDC would be as follows:

- Online monitoring of the scientific payloads data and health parameters.
- Generate necessary feedback and strategies for payload control and navigation.
- Coordinating with potential users.
- Examining the validation and authenticating for scientific use.

---

[5] Based on the structure operational during the moon mission it could be concluded that each POC would be co-located at the respective institutions of the PIs of the various on-board experiments, which will generate the higher-level products depending upon their requirement and various applications. The PI will also coordinate the science to be done with the data with other investigators in a given experiment, who could be from different institutions

[6] Author's visit to the location and discussions with the scientists there

- Associating the attributes of different payload data to composite data group for effective data utilisation.
- Continuously monitoring the history and trends in space and time.
- Correlating the available data with previous Mars missions for validation and new findings.
- Developing new algorithm and techniques for interpreting, analysing and presenting the data based on laboratory simulation of lunar samples and detector characteristics.
- Design of comprehensive chemical and mineralogical map based on all the current and previous missions.

In case of Moon mission, India had some sensors onboard from various other countries and naturally the data inputs were shared by various countries. However, India's Mars mission is an entirely Indian affair, and at present, there are no proposals to share the payload data with other countries. It is expected that subsequently, based on the data analysis, few important research papers would be produced and they would be submitted for the publications in peer reviewed scientific journals.

5.3 Undertaking Challenges

Since time immemorial undertaking developments in technologies have always been challenging and same is the case with space technologies. Over a period of time, various space technologies particularity in the fields of communication, navigation, remote sensing etc. have reached to a certain level of maturity. However, in the fields like deep space much needs to be achieved in regards to technology. For many years, humans have ambition to reach Mars and subsequently to establish human colonies over there. Presently none of the space-faring states are even somewhat close to realise this dream. Mars offers the greatest challenge for the rocket scientists, space scientists, aerospace engineers and astronomers. Also, there are some philosophers, intellectuals and thinkers from all over the world raising certain pertinent sociopolitical questions. All this puts the policy formulator and state administration which always face problems with the budgets into a quandary. Hence, for any state developing the missions, the Mars mission would always have to face challenges at various planes. Such challenges are discussed elsewhere in this book. It is important to note that such challenges essentially involves bringing change in the perceptions and are need to be addressed by the policy makers and the technocrats. The scientists basically have to confront situation in regards to the technological challenges and at times the challenges posed by the nature. This section restricts towards addressing the anticipated technological challenges for the Mars mission.

The success rate in case of various Mars missions undertaken so far is not encouraging (approximately 1 in 3 becomes successful); hence, the scientific community over a period of time has became well aware of the challenges. They have become aware of the fact that strategic knowledge gaps do exit and are trying their bests to overcome them. Also, since ISRO is undertaking this mission without any external assistance they have been involved in developing few subsystems for the first time. Naturally, they would have to be extra careful.

Over the years, it has been observed that the different categories of Mars missions have their own set of challenges. Flyby, orbitor, rover and lander missions have few common challenges while there are also many mission-specific challenges. For a spacecraft to navigate through weather, gravity and radiation for a longer distance and survive for minimum of 1 year and keep the communication alive with the earth stations is challenging. Historically, it has been witnessed that space agencies learn from the successes and misfortunes of previous missions but at times in spite of taking all precautions based on the previous experiences they end up is experiencing some new set of challenges. Hence it is also important to remain prepared to address surprises.

The unmanned missions are expected to encounter various environmental and technological challenges while the human missions could face technical, physical, and psychological challenges. Human Mars missions (when it happens) risks of radiation exposure, de-conditioning and the psychological impacts of isolation. There are very many challenges in devising human missions to Mars. One of the overriding factors that make Mars missions fundamentally different from lunar missions is the fact that there is very little opportunity to abort the mission. This drives up the requirements and the cost of systems, and since Mars missions will tend to be over 2.5 years roundtrip, extensive (and expensive) life-testing will be required [22].

Indian scientific community involved with its Mars agenda is fully aware of all such challenges. They have their own experiences in the deep space area with their Moon mission. This mission was planned to endure for 2 years. This mission was a successful mission and has very many positives associated with it; however, the mission lasted for 9 months and could finish about 95 % of its stated objectives. The early abandonment of this mission deprived ISRO form the collection of more data. ISRO community is also aware about the challenges in regard to Mars orbital insertion for the orbitor. Then, there are challenges associated with power systems, communications and few other issues. They are also aware of the fact that the orbitor would travel around 300 days and subsequently all its systems including the engines need to function without any glitches after experiencing the unknown environmental conditions for some amount of time. Following paragraphs indicates that how ISRO is planning to address some of such challenges.

5.3.1 Power Systems

The power system required to support the mission during various phases of mission like transfer orbit and on-orbit phase. The power system consists of power generation, energy storage and power conditioning elements. One of the major challenges in the design of power system is due to the larger distance from Sun the dependence on solar energy reduces. Mars is the last planet in the solar system where solar power generation can be used effectively. The power generation in Mars orbit is reduced to nearly 50 % compared to Earth's orbit. Due to the eccentricity of Mars orbit around sun, the power generation variation is nearly 15 %. If one Watt is power generation when Earth is at perihelion, 0.35 W will be the power generation when Mars is at aphelion. The ordeal for the scientific community is to cater for such power generation variations at different distances and in different orbits. Also, there are specific requirements for powering the spacecraft during eclipse phase and payload data download phases. ISRO is making specific provisions to cater for such challenges.

5.3.2 Communication Systems

The communication systems for the Mars mission are responsible for the challenging task of communication management at a distance of nearly 200–400 million km. For one way communication (from earth station to Mars orbitor and back), it would take approximately 20-min time. Hence, ISRO would have to cater for 40 min delay in the communications and make the planning accordingly. To cater for such requirements ISRO has invested into the Telemetry, Tracking and Commanding (TTC) systems and Data transmission systems. For this purpose, ISRO has developed a major infrastructure for the antenna system constituting of low, medium and high gain antennas. In fact, a major structure was erected during the Moon mission (2008) and that has been upgraded to cater for the requirements related to Mars.

5.3.3 Ground Segment

To support such a complex mission, it is essential to have a state-of-art ground infrastructure available to realise the operational communications between a control centre and a spacecraft. This being India's second deep space mission luckily; the ground station is in place but in regards to distance as compared to Moon, and Mars is a quantum jump. The Indian Deep Space Network station (IDSN32), a 32-m-diameter network antenna system located in Byalalu, Bangalore, was established with a view to meet, not only the requirements of Chandrayaan-1 mission, but also ISRO's future missions to Mercury, Venus and up to Mars [23]. So, the specifications of IDSN-32 are arrived such that it will be able to cater TTC and science data reception functionalities for a mission to Mars. The hallmark of Indian DSN facility is the state-of-the-art technology 32-m-diameter Beam Wave Guide (BWG) antenna that had been indigenously designed, developed and installed to support all future deep space missions of ISRO. For Mars mission certain amount of mission-specific modifications have been carried out keeping in view the requirements of this mission[7].

To manage a deep space mission for 24 hours, the geometry necessitates the support of minimum two ground stations: one located in eastern and other in western hemispheres. Ground support from JPL-NASA stations is envisaged to support the mission in addition to IDSN support.

The long coasting of PSLV PS4 stage for 1644s before PS4 operation requires two portable sea-borne S-band terminals to be deployed in Pacific Ocean to monitor PS4 and satellite separation.

5.3.4 Propulsion Systems

A spacecraft propulsion system is a method used to accelerate the spacecraft (satellite), and there are different ways to do this with each having some advantages as well as limitations. Reaching Mars could be viewed as one of the most challenging task for the rocket scientists. This mission being first attempt by ISRO to reach such a distance, they are putting significant amount of efforts to develop their systems. Their propulsion systems embody the truly enabling technology for departing earth and reaching the Mars. The Mars Orbiter Mission propulsion system has its heritage from GEO mission and consists of a unified bipropellant system for orbit raising and attitude control. It consists of one 440-N Liquid Engine (LE-440) and 8 numbers of 22-N thrusters. The propellants are stored in Titanium propellant tanks each with a capacity of 390 L pressurised with Helium gas. The tanks have combined storage capacity up to 850 kg propellant. The 67 litre helium pressurant tank is used to pressurise the propellant. The 22-N thrusters are used for attitude control during the various phases of the mission like orbit raising using liquid engine, attitude maintenance, Martian orbit maintenance (if any) and momentum dumping. As the last few LE burns around Mars are to occur after 10 months of launch (this is because it would take 300 days, time to reach Mars), suitable isolation techniques are adopted to prevent fuel/oxidiser migration issues. For around 300 days the craft has to experience the vagaries of space weather including the radiation. The biggest challenge is that after undertaking the journey of 9/10 months the all systems of the orbitor need to restart and function properly.

Similar to conventional GEOSAT missions, Main Engine is proposed to be isolated after the earth-bound Liquid Engine operations are completed. Liquid Engine would be isolated by operating Pyro valves. On completion of coasting phase, the pyro valves would be commanded OPEN and propellant supply to Liquid Engine would be re-established for Martian Orbit Insertion (MOI) manoeuvres.

---

[7] Author's visit to ISRO's Byalalu, Bangalore facility and discussions with the scientists' onsite

5.3.5 On-board Autonomy

On-board autonomy refers to the capacity of the orbiter to make its own decisions about its actions. As the distance between the Mars Orbiter and earth increases, the need for autonomy increases dramatically. Given an average round trip time (to and fro) from earth to Mars of approximately 40 min, it would be impractical to micromanage a mission from Earth. Due to this communications delay, mission support personnel on Earth cannot easily monitor and control all the spacecraft systems in real-time basis. Therefore, it is configured to use on-board autonomy to automatically manage the nominal and non-nominal scenarios on-board the spacecraft. Autonomy is in charge of the spacecraft when communication interruptions occur when the spacecraft it occulted by planet Mars or sun. Autonomy also ensures the recovery from safe mode occurrences on-board the spacecraft.

5.3.6 Weather

It has been emphasised that for Mars Orbitor Martian atmosphere, real-time climate conditions and radiation could pose some challenges. ISRO needs also to remain prepared to handle the challenges of bad weather (if any) on ground and extending to stratosphere (50/60 km above the earth's surface), during the launch phase. From the suitability of weather point of view, month of November is not the most appropriate time to plan the launch. However, because of the astronomical compulsions ISRO would have no options but to abide by the available launch window.

Mars mission would be launched from ISRO's Satish Dhawan Space Centre which is situated at Sriharikota in Andhra Pradesh. This location is 80 km (50 mi) from the city of Chennai. During the month of November, this entire region is in the grip of either one of the two major rain generating seasonal weather systems. This region is the part of Indian peninsula and weather over here could get affected either by the south-west Monsoon (during its delayed withdrawal phase) or by active north-east Monsoon.

The climate averages for the region indicates that from the peninsular India normally the south-west monson withdraws by mid to end of October every year. Certain amount of variations could take place with the withdrawal pattern and at time it has been found that withdrawal does gets delayed. On the other hand, the normal date of onset of the north-east monsoon is around 20 October with a deviation of about a week on either side. The rains in this season continue till December. The major period of rainfall activity is over south peninsula, particularly in Coastal Andhra Pradesh, Rayalaseema and Tamilnadu–Pondicherry region. For Tamilnadu, this is the main rainy season accounting for almost half of the annual rainfall [24].

In addition to the monsoon rains, month of November is also the period when Bay of Bengal and adjoining costal region is most susceptible for cyclonic storms (hurricanes) formation [25]. Probability of occurrence of the cyclonic storms is maximum during this period. All this clearly indicates that ISRO has to remain prepared to guard this rocket from adverse weather prior to and during launch phase. They have to guard against the weather hazards like heavy rains, intense clouding, thunderstorms/lightning/hail and strong wind conditions with every hazard having potential to cause damage to the rocket.

The Sriharikota High Altitude Range (SHAR), Satish Dhawan Space Centre, Sriharikota launch complex is qualified for all the weather conditions throughout the year. ISRO has the history of launching from till the date October 22 and no launches have ever taken place in November. SHAR is equipped with advanced weather monitoring systems and a dedicated team is working on the weather analysis. In case of an unsuitable weather forecast for the launch date, the mission is planned in such a way that the launch can be postponed or advanced by one day without any penalties. This means that ISRO has a system in place to cater for bad weather. What is important in case of Mars mission is to note that till date ISRO had avoided in undertaking any launch form the India soil in November/December. Even during the month of October, they have undertaken few launches and most of them being before 15th October. In October 2008, India's moon mission was launched on 22nd of the month [26]. The main challenge in the month of November is that the weather conductions during active monsoon phase or during the passage of a cyclonic storm normally remains bad for 3–5 days at a stretch.

5.3.7 Comet Strike

When India's mission is its final stages of preparations, suddenly a new challenge has arisen and this is not a technological challenge, but a challenge of entirely different nature which normally no one could have even imagined to occur.

Very recently NASA has predicted that there is an outside chance that a newly discovered comet might be on a collision course with Mars. Astronomers are still determining the trajectory of the comet, named C/2013 A1 (Siding Spring), but at the very least, it is going to come fairly close to the Red Planet in October of 2014 [27]. The nucleus of the comet is probably 1–3 km in diameter, and it is approaching fast, around 56 km/s (125,000 mph) speed. If it does hit Mars, then it is likely to deliver as much energy as 35 million megatons of TNT (may be leave a crater about 500 km wide and 2 km deep). For comparison, the asteroid strike that ended the dinosaurs on Earth 65 million years ago was about three times as powerful, 100 million megatons. Another point of comparison is the meteor that exploded over Chelyabinsk, Russia, in February of 2013, damaging buildings and knocking people down. The Mars comet is packing 80 million times more energy than that relatively puny meteor [28]. Initially, the scientists were putting the odds of likely impact at 1 in 2000 and subsequently to about 1 in 8,000. However, based on Apr 7, 2013, data set it has been calculated that the chances of the comet impacting the Red Planet are about 1 in 120,000 [29]. This comet which was only discovered on Jan 3, 2013 was posing a major challenge for the India decision makers. What was at stake was the Indian investments of about US$ 83 million. But now the recent assessment must have given a breather to India's mission managers!

Just in case such a low probability threat turns out to be a reality, then it could have a major impact on various existing (Opportunity and Curiosity rovers) and various future Mars missions particularly the one which are likely to takeoff in 2013. A big impact would lift a lot of *"earth"* into the Martian atmosphere–dust, sand and other debris. India's mission likely for November 2013 launch is expected to take 9 months to reach Mars around September 2014 and the estimated time for close approach to Mars for the comet is around Oct 19, 2014. The probability of the comet hitting the Mars is very less, but there is another factor which looks worrisome and that is the possibility that Mars will be engulfed in the tail of the comet—extending to millions of kilometres—which will be on Mars' sunward side.

For India's orbitor or for any other craft viewing Mars, there is no guarantee that they could get a clear view of planet because of the tail of the comet. Four decades back, NASA's Mariner 9 which had arrived at Mars on November 14, 1971 was not able to take any pictures for 45 days due the Martian surface being engulfed by a huge dust storm. What is more problematic is not only the visibility of the planet but about the composition of the tail. The comet tails are usually the streams of dust and gas. Most comets have presence of methane in their tails. One of the important scientific objectives of the India's mission is assessment in regards to methane in Martian atmosphere. Amongst India's five payloads, one is the methane sensor for Mars (MSN). There is a good chance that this MSN payload may confuse the methane it detects from the comet as that of Mars and transmit wrong data. Such data could be misleading [30]. This knowledge of the likely presence of the comet close to Mars offers a challenge to the ISRO's scientific community and they would have to factor this possibility in their mission planning.

5.3.8 Overall Mission Intricacies

India's mission could be viewed as simple mission in comparison with the most of other missions launched so far. However, it is important to note that since globally very few missions have succeeded so far, hence many scientists all over the word are keenly monitoring Indian investment in the Mars project [31]. They feel that there is much to learn from the way India has devised this project and are keenly looking forward for the information which could be made available.

For ISRO, Mars mission is a technology demonstrator till the craft successfully reaches Mars, but after that the mission would get converted into a science mission. Hence, there could be different set of challenges when the mission reaches Mars. There would be five different payloads undertaking various observations. Every payload has a specific role. There exists a possibility some unknown technological challenges from sensor malfunction to communication breakdown could emerge at any point in time. Naturally, ISRO would have to remain prepared for any eventualities because challenges do have habit to come unannounced.

Acknowledgments Author is grateful to ISRO for sharing the information which has been used for the organization of this chapter. This information also includes various photographs which are used for the generation of figures and table in this chapter.

References

1. Sarathi VP. Ancient Indian mathematics and Astronomy. http://www.indicstudies.us/Astronomy/aimword.pdf. Accessed Apr 10, 2013.
2. Boonrucksar S. Astronomy in Asia. In: Hearnshaw JB, Martinez P editors. Astronomy for the developing world, International Astronomical Union, Cambridge: Cambridge University Press; p. 117.
3. Inter-University Centre for Astronomy and Astrophysics (IUCAA). http://www.iucaa.ernet.in/Mission.html. Accessed Apr 10, 2013.
4. http://www.astron-soc.in/index.html. Accessed Apr 8, 2013.
5. http://www.prl.res.in/. Accessed Apr 7, 2013.
6. http://www.iist.ac.in/. Accessed Apr 10, 2013.
7. ISRO studying proposal on mission to Mars, the Times of India, Apr 11, 2007 and http://news.oneindia.in/2008/11/23/mission-to-mars-next-ambition-of-isro-madhavan-nair-1227412909.html, November 23, 2008. Accessed Apr 11, 2013.
8. Indo-US Space accord paves way for future joint exploration, Mar 16, 2006. http://www.spacedaily.com/reports/Indo_US_Space_Accord_Paves_Way_For_Future_Joint_Exploration.html. Accessed Apr 8, 2013.
9. HAL delivers Mars orbiter mission satellite structure to ISRO. http://www.thehindubusinessline.com/news/science/hal-delivers-mars-orbiter-mission-satellite-structure-to-isro/article3922481.ece. Accessed Jan 18, 2013.
10. Bagla P. India's maiden mission to Mars later this year, Jan 7, 2013. http://www.ndtv.com/article/india/india-plans-to-launch-mars-mission-later-this-year-314208. Accessed May 3, 2013.
11. http://www.isro.org/launchvehicles/launchvehicles.aspx. Accessed Apr 02, 2013.
12. http://www.esa.int/Our_Activities/Operations/What_are_Lagrange_points. Accessed May 2, 2013.
13. http://mars.jpl.nasa.gov/programmissions/missions/missiontypes/orbiters/. Accessed Apr 15, 2013.
14. Wallace TF. An introduction to space mission planning. http://design.ae.utexas.edu/mission_planning/mission_resources/mission_planning/Intro_to_Mission_Planning.pdf.   Accessed Apr 28, 2013.
15. http://www2.jpl.nasa.gov/basics/bsf4-1.php. Accessed Apr 28, 2013.
16. Flory JA, Kharoufeh JP. Optimal satellite payload selection and specification. Mil Oper Res. 2010;15(3):43–5.
17. http://www.isro.org/pdf/Annual%20Report%202012-13.pdf. p 69, Accessed Mar 30, 2013 and based on the information received for ISRO.

18. http://www.isro.org/isrocentres/bangalore_electroopticlab.aspx. Accessed Jan 24, 2013.
19. http://www.sac.gov.in/SACSITE/asac-anoverview.html. Accessed May 2, 2013.
20. http://www.vssc.gov.in/internet/getPage.php?page=Projects%20Page&pageId=221. Accessed May 2, 2013.
21. Ramachandran R. Frontline. 2008;25(24).
22. Hawley SA. Mission to Mars: risks, challenges, sacrifices and privileges: one Astronaut's perspective. J Cosmol. 2010;12:3517.
23. Hathwar GR. IDSN story. ISRO, Bangalore: Publications and Public Relations; 2009.
24. www.imdchennai.gov.in/northeast_monsoon.htm. Accessed Apr 14, 2013.
25. Singh OP et al. Has the frequency of intense tropical cyclones increased in the north Indian Ocean?. Current Sci. 2001;80(4)575.
26. http://www.isro.org/satellites/allsatellites.aspx. Accessed Apr 27, 2013.
27. http://science.nbcnews.com/_news/2013/02/26/17107085-comet-just-might-hit-mars-in-2014?lite. Accessed Mar 12, 2013.
28. http://www.spacedaily.com/reports/Collision_Course_A_Comet_Heads_for_Mars_999.html. Accessed Apr 14, 2013.
29. http://www.spacedaily.com/reports/Comet_to_Make_Close_Flyby_of_Red_Planet_in_October_2014_999.html. Accessed Apr 15, 2013.
30. http://www.dnaindia.com/scitech/1819996/report-dna-exclusive-comet-mars-isro-s-rs450cr-dream-mission.
31. Bagla P. Mars mission: demonstrating India's technology. http://www.bbc.co.uk/hindi/science/2012/08/120803_mars_mission_ind_fma.shtml. Accessed Apr 12, 2013.

Chapter 6
Asia's Investments in Mars

> By its very size, big science cannot survive in isolation from the
> nonscientific spheres of society. It has become an economic,
> political, and sociological entity in its own right.

Asian mythology could be viewed as an important element of its culture, which
allows having glimpse into its thinking in anent era about the solar system. Local
wisdom in many Asian countries reflects their interest in astronomy since the
historical period [1].

The most interesting aspect of Asian mythology is that even though at times, it
is found projecting few astrological superstitions but still it broadly presents more
of a scientific and philosophical approach. Hence, more than displaying myths
about the stars and planets it is found providing some rationale for their existence
and motions. Some commonalties as well as differences could be found among the
Asian and Roman mythologies. There are different narratives explaining the
presence of planets and the universe as such but there has always been an urge to
learn more about the planets. "Deciphering Mars" appears to be an area of special
interest both in Roman and Asian mythologies. Normally, it has been observed that
there is certain amount of association of Mars with the wars in various narratives.
Ares was the Greek god of war and the counterpart of Ares among Roman gods is
Mars. This was the most worshiped military god then. The annual celebrations
associated with Mars were held in month of March, a month named after him. The
Mesopotamians had named this planet as Nergal.

Ancient Chinese astronomers named the planet Mars as "Ying Huo" the "Fire
Star". Both the Chinese and Korean cultures refer this planet by the same name
which is actually derived from the ancient Chinese mythological cycle of five
elements. According to ancient Chinese mythology, Mars is personified as a
powerful red-faced warrior, living at the east end of heaven causing many wars.
The planet was known by the ancient Egyptians as "Horus of the Horizon"/"Horus
the Red" or Red-Horus (the "planet smasher"). In India (Hindu/Sanskrit languages), the Mars is called Mangala and other names are Angaraka and Kuja.

*Peter Galison, "The Many Faces of Big Science", In Big Science: The Growth of Large-Scale
Research. Peter Galison and Bruce Hevly, Eds. (Stanford, CA: Stanford University Press,
1992). This source has been quoted by James A. Vedda (2008), "Challenges To The
Sustainability of Space Exploration", *Astropolitics: The International Journal of Space Politics
& Policy*, 6:1, p. 22.

Angaraka means the one who is red in colour. Mars is also known by name mostly used by the Indian astrologers called Bhauma (meaning son of Bhumi) in Sanskrit. He is known as the god of war and is celibate [2]. Mars was considered by various civilianisations as a violent god and associated with the disruptions on the Earth. Egyptian and Chinese civilisations in yesteryears had called Mars as a wolf star. Chinese records mention a time when two planets battled in the sky and they were Venus and Mars [3, 4]. For every civilisation, there have been certain beliefs associated with the planet Mars. Broadly, the primeval connect of Mars with society has been more about wars and damage. Mars has also been considered as an unpredictable planet and probably that could be the reason for the humans to continue with their quest to learn more about Mars for all these centuries!

After the beginning of satellite age, the interest in Mars was found growing globally and Asia was no exception. However, western states were the early starters in respect of launching satellites/probes for Mars missions. In Asia, so far only three states, namely Japan, China and India, have shown interest to learn more about Mars by planning dedicated missions to Mars. During the Cold War era, the interests of Asian states in Mars were found rather limited. Limitations of technology and the financial concerns appear to be the main reasons behind this. However, with increase in space activities and appreciating the techno-economic importance of Mars, some of the Asian space powers are expected to improve their investments in near future. Also, it is quite understandable that since only Japan–China–India in Asia has reached to a certain level of expertise in space area, they would only plan to assume such missions. Another space-faring state in Asia, the Israel also has the potential to undertake activities like the Mars mission. It has been reported that the cooler used on-board Curiosity rover was manufactured in Israel's company Kibbutz Ein Harod. It is meant for identifying Mars materials. The requirement is to keep the detector at -173 °C at all times, and this particular cooler has been used for this purpose [5]. However, in spite of having a strong industrial base, Israel is not keen to make investment in the Mars. The state has officially announced that they would not be venturing to Mars at least in near future (this was announced after the landing of NASA's Curiosity rover on Mars).

Following portion of this chapter discusses attempts by Japan and China towards exploration of Mars.

6.1 Japan's Mars Agenda

The first mission towards Mars was undertaken by the erstwhile USSR in 1960, while in Asia first state to attempt a Mars mission was Japan and this mission was launched in 1998, almost four decades after the first attempt in the world to reach Mars. This was Japan's first interplanetary mission named Nozomi (PLANET-B) meaning Hope. This mission was aimed to orbit a satellite in the close vicinity of Mars and to transmit data for one Martian year. It had fourteen instruments on board including sensors from EU, Canada and the USA. Initially, the performance of the craft was found normal in the Earth–Moon system. On the way to Mars, however, troubles occurred and substantial orbit changes were made. Thus, overall it took 4 more years than the original plan and the mission approached closely towards Mars by December 2003, however, Mars orbit insertion could not be achieved and the craft was lost [6].

This mission was launched on Jul 4, 1998, and it was scheduled to arrive at Mars on Oct 11, 1999. After launch, the Nozomi was put into an elliptical geo-centric parking orbit with a perigee of 340 km and an apogee of 400,000 km. As per the procedure, Nozomi flew two Lunar swingbys[1] on September 24 and on Dec 18, 1998, and one Earth swingby on Dec 20, 1998 and was put into an escape trajectory towards Mars. But the Earth swingby left the spacecraft with insufficient acceleration and two course correction burns carried out on 21 December used more propellant than planned, leaving the spacecraft short of fuel. Basically due to a problem in the propulsion system, the craft got "insufficient acceleration" and was not able to achieve the required orbit. However, the mission was not abounded. A new plan was devised and as per that Nozomi was to remain in heliocentric orbit for an additional 4 years and encounter Mars at a slower relative velocity in December 2003. The rescheduled orbital insertion was to take place on Dec 14, 2003. However, few problems cropped up during this period. The electric system of Nozomi as well as the S-band antenna suffered damage from a heavy solar eruption in April, 2002. This restricted the communication possibilities with the spacecraft. Finally, a correction manoeuvre on Dec 9, 2003 failed and this prompted the Japan's space agency (JAXA) to abandon the mission [7].

Subsequently, Japan has not officially announced any other major plans for visiting Mars. However, there is some information available which explains a new proposal for the future. The new Mars mission is named Mars Exploration with a Lander and Obiter (MELOS). For this mission, Japan's H-IIA rocket (with 4 solid propellant boosters) would be used to put up 3 tons of spacecraft into Mars orbit. MELOS is expected to be launched during the window in 2018 and the likely key focus of this mission could be to study Mars meteorology [8]. The mission is also planning to undertake seismic study and could attempt to study the process of evolution of Mars. However, presently not many deliberations involving Mars mission are found taking place within the state.

Apart from Mars, Japan is also perusing an agenda of other planets and asteroids. Japan unsuccessfully attempted a mission to Venus during December 2010. Japan's spacecraft Akatsuki ("dawn" in Japanese) failed to inject the orbiter into the planned orbit as a result of incorrect orbit estimation. They propose to undertake a renewed attempt to get it into orbit around Venus in 2015 [9]. Japan has undertaken a partially successful Hayabusa mission to an asteroid. It was a sample return mission which was launched during 2003 and returned to Earth by 2010 however, it had failed brought back the samples as proposed. Japan is

---

[1] Apogee is the point where an orbiting body is at its farthest from the Earth. An interplanetary mission in which a space vehicle uses planetary gravitation for changes in course.

expected to launch Hayabusa2 in 2014. It appears that Japan wants to build on the success of its first asteroid mission.

When asked to offer his views on the Japan's overall deep space agenda, the Prof Kazuto Suzuki[2] mentioned that "Japanese space agency, particularly its science research arm, ISAS, chose exploration programme based on their internal distribution of resources. Mars is a mandate for the planetary science group, which is different from asteroid sample return programme such as Hayabusa. The next step for planetary science is 'BepiColombo'[3] [10], Mercury programme with ESA. The turn of Mars will be a bit far away for the next round. Hayabusa has attracted social and political attention, so it has been given a higher priority, but it is considered as 'engineering testing programme' rather than scientific exploration programme. Generally, there is less enthusiasm in Japan for Mars exploration. Moon is much popular celestial body".

6.2 China's Interests in Mars

China has got keen interest in the business of interplanetary space science. China's Information Office of the State Council has published a white paper on China's space activities in 2011 outlining China's plans for investments in the space area. This white paper mentions that "China will conduct special project demonstration in deep space exploration, and push forward its exploration of planets, asteroids and the Sun of the solar system".

On Mar 26, 2007, China announced its first international cooperative project of a joint Chinese–Russian exploration of Mars. Both the sides signed the cooperative agreement towards that effect with the Chinese and Russian heads of state as witnesses [11]. Unfortunately, this China's first planetary mission called Yinghuo-1 Mars orbiter failed due to the launcher and orbit burn failure. Yinghuo-1 was launched along with the Russian craft called Fobos-Grunt to Mars by a Ukrainian rocket. This small probe was weighing 115 kg and had a designed life for a two-year mission. The main goal of this mission was to search for the signs of liquid water on the Mars surface. The other scientific objectives of this mission included an investigation of the plasma environment and magnetic field study of Martian ion escape processes and possible mechanisms. Overall, the purpose was to explore the space weather of the Mars and test the deep space communication and navigation techniques [12]. This probe had five payloads on board. In the vicinity of Mars, Yinghuo-1 was to be inserted into a near-equatorial, elliptical orbit.

---

[2]   He is the professor of International Politics at Hokkaido University, Japan. He has been closely involved in the development of Japanese space decision-making process for long. He is a Chairman of the Space Security Committee of the International Astronautical Federation.

[3]   Is one of ESA's cornerstone missions, it will study and understand the composition, geophysics, atmosphere, magnetosphere and history of Mercury, the least explored planet in the inner Solar System.

The spacecraft was to approach within 400 to 1,000 km of Mars' surface at closest approach. The mission was tasked to undertake various measurements. Phobos-Grunt and Yinghuo-1 were also expected to work in tandem to measure the structure of Mars' ionosphere [13].

The entire Sino-Russian mission had a great promise however due to the launcher failure these probes missed an opportunity to prove their mettle. The Ukrainian Zenit rocket was launched on Nov 8, 2011, with Yinghuo-1 and the Russian Fobos-Grunt spacecraft onboard. Fobos-Grunt was to perform two burns and orbit-raising manoeuvre two and a half hours after launch [14]. This was meant to depart Earth orbit and begin its journey to Mars. However, these burns did not take place. Further efforts to retrieve the situation failed and on Nov 17, 2011, China formally declared the loss of Yinghuo-1 probe.

It is important to appreciate that the Yinghuo-1 probe never got an opportunity to prove its worth hence to call Yinghuo-1 mission a failure would be unfair. The entire process of the developing the mission has definitely provided the Chinese scientific community a considerable amount of experience. It was expected to be a fairly low-cost mission and groundwork did exposed Chinese engineers towards designing and assembling a flight-ready spacecraft. Also, various other communication mechanisms were developed as a part of mission activity. Even though the mission could not go through, now China has a structure available in respect of procedures for controlling the spacecraft and analysing its data [15]. For this mission, Chinese had entered into international partnerships for acquiring certain instrumentation facilities/hardware. The overall knowledge gathered during the entire process of development of this mission could allow them considerable assistance for indigenisation of technology in this field. All this is going to assist them considerably for future Mars missions.

There is no clarity that whether China would use the next "launch window" in 2013 or 2016. China is presently working towards catching the earliest possible launch date. In 2013, China may launch its solo Mars mission by using their own rocket system (best option is probably Long March 3B), observation device and detector. This Mars probe is expected to be an updated and modified version of lunar probe [16]. However, there is no official confirmation in this regard.

As per Prof Wu, Riqiang[4], "It seems that China will launch a small satellite to Mars this year (2013)". However, China has not yet released its future plan in this regard. He feels that there are no substantial benefits from Mars programme and the purpose behind these investments is mainly for the reputation of the state followed by technology development and assistance in developing some infrastructure for deep space exploration. He does not view any significant glamour associated with Mars programme among the Chinese population and feels that Chinese people are just proud that their country can undertake such mission like the United States/Russia/India. He further added that, no one within the state are

---

[4] He is an Associate Professor at School of International Studies at Renmin University of China and expert on space issues.

found debating the need for such missions and there are no concerns in respect of the cost aspects.

Another Chinese scholar Prof LI Juqian[5] is of the opinion that the technology challenges regarding Mars exploration also could have an impact in the scheduling of such missions. Presently, there is no specific and clear project on Mars exploration. He mentioned that "some Chinese scientists feel that; it might need about twenty more years for China to explore the Mars independently". He also feels that, "exploration of Mars is a part of exploration of the solar system; so it is not going to be just one time attempt".

Overall, it could argued that after the mishap with the Yinghuo-1 probe, China is steadily perusing it Mars agenda and is not found in any great hurry. China appears to have long-term interest to study Mars seriously. China's third Moon mission Chang'e 3 is expected to be launched during second half of 2013. This mission is scheduled to perform soft landing on the Moon after the gap of almost 37 years. The mission is likely to be beneficial for China's futuristic human Moon programme. Naturally, more concentration is likely to happen on this programme.

Mr Ouyang Ziyuan, chief scientist of China's lunar orbiter project, is of the opinion that a three-phase probe to Mars could be envisaged. The three stages of the Mars probe will entail remote sensing, soft-landing and exploration, and the probe would return after automatic sampling. He feels that the Moon-landing orbiter Chang'e-3 could help build a telecommunication network that covers a future Mars probe [17]. China appears to be working on various allied aspects in respect of reaching and experimenting on Mars. Chinese researchers have completed the country's first test of a self contained system that could grow vegetables on the Moon or Mars [18]. China was also part of the Mars-500 experiment (2007–2011) [19] aimed at obtaining experimental data on the health of astronauts and their ability for work in situation of prolonged isolation. It was joint mission undertaken by Russia, Europe and China. For this purpose, a 520-day manned Mars mission was simulated where six astronauts were kept in isolation for all these days. This group had one astronaut from China. This experiment has yielded valuable psychological and medical data. Such participation would assist them in their proposed manned Moon/Mars mission.

6.3 Chinese and Japanese Mission Payloads

In comparison with the China's Yinghuo-1 mission, the only other Asian state to launch a Mars mission was Japan (the unsuccessful mission Nozomi/Planet-B) and their mission was launched almost 15 years earlier. Naturally, the status of

[5] He is a law professor at China University of Political Science and Law, and Council-Member of the Space Security Council of World Economic Forum and Standing Council-Member of the China Institute of Space Law.

technology then could have been much different, and hence, it would be unwise to undertake any form of strict technology comparison in regard to these two missions. However, for the sake of broad understand about the scientific intent of the Mars programmes of these two states, some broad idea could be congregated by knowing the nature of payloads used by them onboard of these missions.

The main purpose behind Japan's Mars mission was to research Martian upper atmosphere by focusing on the nature of interaction with solar wind. The aim was to conduct research on interaction between Martian upper atmosphere and solar wind, undertake observations of Martian magnetic field, carry out the remote sensing of Martian surface and satellites, etc. To fulfil all such research objects following scientific instruments [20] were put onboard of Nozomi.

1. Mars Imaging Camera (MIC)
2. Magnetic Field Measurement (MGF)
3. Probe for Electron Temperature (PET)
4. Electron Spectrum Analyser (ESA)
5. Ion Spectrum Analyser (ISA)
6. Electron and Ion Spectrometer (EIS)
7. Extra Ultraviolet Scanner (XUV)
8. Ultraviolet Imaging Spectrometer (UVS)
9. Plasma Wave and Sounder (PWS)
10. Low Frequency Plasma Wave Analyser (LFA)
11. Ion Mass Imager (IMI)
12. Mars Dust Counter (MDC)
13. Neutral Mass Spectrometer (NMS)
14. Thermal Plasma Analyser (TPA).

Chinese mission had less number of payloads in comparison with Japan. This mission had two agendas: one, it had its own defined tasks; and second, it was to operate jointly with the Russian craft Fobos-Grunt. Overall, the Chinese mission had various interesting attributes. It was a fairly low-cost and low-key mission a perfect platform for China to begin its planetary exploration agenda. For this mission, China had also forged some international partnerships for designing/acquiring instrumentations.

Yinghuo-1's primary scientific objectives were to conduct detailed investigation of the plasma environment and magnetic field, study Martian ion escape processes and their possible mechanisms and observe sandstorms. They were also supposed to conduct ionosphere occultation measure with the Russian probe by coordinated observations of the Martian space environment. To undertake various activities, China had four primary scientific payloads on board their craft:

1. A plasma package, consisting of an electron analyser, ion analyser and mass spectrometer
2. A fluxgate magnetometer
3. A radio occultation sounder

4. An optical imaging system, consisting of two cameras with 200 m (660 ft) resolution, allowing high-quality images of the Martian surface to be captured from orbit [21, 22].

It is important to note that even in their failures with the Mars missions both these states have gained significant experiences. The Japanese mission was aborted because the spacecraft failed in its injection into orbit around Mars due to unrecoverable malfunction. However, this mission remained in the space for an approximate 5-year period in heliocentric orbit. Hence, it was possible to undertake significant amount of observations. The Mars imaging camera (MIC) has taken some pictures too.

One of the sensors, the Ultraviolet Imaging Spectrometer (UVS) successfully observed the hydrogen Lyman-alpha line (this is a spectral line of hydrogen/on-electron ions in Lyman series) in interplanetary space. The distribution map of Lyman-alpha observation enables detailed research into the property of solar wind. Extreme Ultraviolet Telescope (XUV) was to be used to capture the charged particles (plasma) which might resemble the change in magnetic field by the presence and movement of helium ion. It has been revealed by Nozomi observation that much more volume of helium ion than expected is streaming out from the region observed. Another sensor the Dust Counter (MDC) had began its observation just after the launch and had continued to observe the velocity and mass of dusts around Earth and interplanetary space. This detection of dusts of extra-solar system origin in near-Earth region is one of the most important results done by Nozomi. The Ion Energy Analyser (ISA) had succeeded in detecting plasma which was part of solar wind reflected by the Moon. Electronic Energy Analyser (ESA) had helped in the detection of electrons. High-Energy Particle Instrument (EIS), together with Ion Mass Imager (IMI), worked as a long-term valuable monitor of observing solar wind functioned as a far distant observation station. Magnetic Flux Instrument (MGF), which had been to observe the magnetic field of Mars, observed the magnetic field of solar wind in interplanetary space [23]. Overall, Nozomi was installed with globally highest-level observation instruments mainly for analysis of magnetic field and plasma. Overall whatever role, whatever location and whatever time, it was asked to observe, analyse and report it did a wonderful job.

China's Yinghuo-1 had payloads like ion and electron. The electron analyser was one of the analysers of plasma package for Mars exploration. It was to analyse the distribution of electron's energy and direction around the orbit of spacecraft. Because the electrons in Martian magnetosphere are strongly isotropic, the electron analyser was to mainly perform 2D detection in ecliptic [24]. The rest of the scientific payload included a plasma detector package and a fluxgate magnetometer for analysing in situ energetic particles and magnetic fields, as well as a small (1.5-kg) camera for high-resolution pictures of Martin surface [25]. Another instrument onboard was radio occultation sounder. The instrument (antenna + receiver) is designed to provide accurate vertical profiles of temperature, pressure and humidity of atmosphere in order to study the climatology and meteorology of the Earth and it is based on the available GPS signal passing through the low atmosphere layers [26]. Unfortunately, these specifically designed equipments on board did not have an opportunity prove their worth.

In general, the scientific objectives of both these missions had some commonalities, and focus was essentially to know more about the Mars environment. The objective was to undertake direct and indirect observations and to draw inferences. Japan had designed an interesting mission and would have definitely added to the existing knowledge of Mars then, if had succeeded. It was comparatively advanced mission (even by present standards) and was critically designed to observe atmosphere at higher level. It would be unfair to claim that Yinghuo-1 failed. Rather, it never even got a chance to unfurl its solar panels or perform any real operations in space.

6.4 Sum Total

Manned Mars exploration could be viewed as an ultimate dream for major space powers. Japan and China fully understand that they would have to greatly accelerate their rate of progress in the space area to have a "meaningful: unmanned or manned" interaction with the Mars. There is a possibility that the rate of overall global failures in conquering the Mars and their own earlier failures in this arena could make them to approach Mars with caution. Also, both these states appear to have their plate full with various other ongoing and planed activities in the outer space. Presently, it appears that in their list of priorities Mars is not on the forefront but at the same time they are not obliterating Mars from their overall space agenda too.

Mars challenge is more of a technological challenge and would essentially lead to developing the science of astronomy. Few other scientific and technological benefits have also been envisaged for future Mars missions. However, it is too premature to search for significant "strategic" benefits with Mars agendas of various nation-states. Mars/Moon missions (manned or unmanned) will always have the subtext of "nationalism" at the backdrop. The geopolitical discourse of tomorrow could use "mission to Mars" as a factor to judge the nation's stature. Naturally, no state would not like to loose on such indirect but valuable benefits and hence particularly China is unlikely to disregard the importance of Mars. Finally, these two important Asian space powers are keen to learn more about Mars and understand the significance of commissioning Mars missions, but appear to be still in the process of undertaking a thorough cost-benefit analysis before committing themselves towards making into any major (financial/technological/ political) investments in this field.

References

1. Boonrucksar S. Astronomy in Asia. In: Hearnshaw JB, Martinez P, editors. *Astronomy for the developing world, IAU Special Session no. 5, 2006*, International Astronomical Union Publication (2007), Abstract.
2. http://www.abovetopsecret.com/forum/thread609525/pg1. Accessed on Mar 2, 2013.
3. Sharon RV. Aba, the glory and the torment: the life of Dr. Immanuel Velikovsky. London: Paradigma Ltd; 1995. p. 229.
4. Birrell A. Chinese mythology: an introduction. Maryland: The John Hopkins University Press; 1993. p. 50–3.
5. Israel has no plans for mars mission, http://www.jspace.com/news/articles/israel-has-no-plans-for-mars-mission/10209, Aug 8, 2012 and http://www.ynetnews.com/articles/0,7340,L-4267605,00.htmland   http://www.israeldefense.com/?CategoryID=483&ArticleID=1584. Accessed on Feb 12, 2013.
6. Harvey B, Smid H, Pirard T. Emerging space powers. Christine: Springer/Praxis; 2010. p. 54–57, http://www.stp.isas.jaxa.jp/nozomi/index-e.html.
7. http://spider.seds.org/spider/Mars/nozomi.html and http://www.solarviews.com/eng/nozomi. htm. Accessed on Dec 26, 2012.
8. Satoh T, WG M. MELOS-Japan's Mars Exploration Plan for 2010s. Geophysical Research Abstracts, vol. 11, EGU2009-14081-1, EGU General Assembly 2009.
9. Than K. Japan probe missed Venus—will try again in six years, Dec 8, 2010, http://news. nationalgeographic.com/news/2010/12/101208-japan-venus-spacecraft-akatsuki-missed-orbit-science-space/. Accessed on Jul 24, 2011.
10. http://www.esa.int/Our_Activities/Space_Science/BepiColombo_overview2. Accessed on Apr 3, 2013.
11. China and Russia join hands to explore Mars, May 30, 2007, http://english.peopledaily.com. cn/200705/30/eng20070530_379330.html. Accessed on Dec 12, 2012.
12. Ping JS, Qian ZH, et al. Brief introduction about Chinese Martian mission Yinghuo-1. In: 41st lunar and planetary science conference, Lunar and Planetary Institute, Texas, p. 1060, Mar 1–5, 2010.
13. China's Yinghuo-1 Mars, Sep 9, 2010, Orbiter http://www.planetary.org/blog/article/0000 2655/linkinghub.elsevier.com/retrieve/pii/linkinghub.elsevier.com/retrieve/pii/ S0275106210000172. Accessed on Sep 9, 2011. The information in this blog is based two different papers written by Chinese scientists in Chinese journals. There are some variations (mostly minor) in the information provided in these papers.
14. Bergin C. Fobos-Grunt ends its misery via re-entry", Jan 15, 2012. http://www.nasaspace flight.com/2012/01/fobus-grunt-ends-its-misery-via-re-entry/. Accessed on Feb 14, 2012.
15. Jones M. Yinghuo was worth it, http://www.spacedaily.com/reports/Yinghuo_Was_Worth_ It_999.html. Accessed on Jan 24, 2013.
16. China likely to launch first probe to explore Mars' surface in 2013, Mar 03, 2011, http:// news.xinhuanet.com/english2010/china/2011-03/02/c_13757750.htm, and Jones M. China goes to Mars, http://www.spacedaily.com/reports/China_Goes_To_Mars_999.html. Accessed on May 15, 2011.
17. China sets sights on collecting samples from Mars, Oct 10, 2012, http://www.reuters.com/ article/2012/10/10/us-china-mars-idUSBRE89917P20121010. Accessed on Feb 1, 2013.
18. Prigg M. Marrows on Mars: China reveals plans to grow vegetables in extra-terrestrial bases, Dec 4, 2012, http://www.dailymail.co.uk/sciencetech/article-2242792/Marrows-Mars-China-reveals-plans-grow-vegetables-extra-terrestrial-bases.html. Accessed on Jan 2, 2013.
19. Details of this experiment are available on http://mars500.imbp.ru/en/index_e.html. Accessed on Jan 2, 2012.
20. http://www.isas.jaxa.jp/e/enterp/missions/nozomi/index.shtml. Accessed on Jan 24, 2013.
21. Jones M. Yinghuo was worth it, Nov 17, 2011, http://www.spacedaily.com/reports/ Yinghuo_Was_Worth_It_999.html. Accessed on Nov 26, 2012.

22. Wang C. Introduction to YH-1, the First Chinese Mars Orbiter, http://www.agu.org/meetings/chapman/2008/acall/Prelim-Program.pdf and http://nssdc.gsfc.nasa.gov/nmc/spacecraft Display.do?id=YINGHUO-1 and http://english.peopledaily.com.cn/200705/30/eng20070 530_379330.html. Accessed on Mar 2, 2013.
23. What "NOZOMI" left with us, http://www.isas.jaxa.jp/e/enterp/missions/nozomi/status_01. shtml. Accessed on Mar 17, 2013.
24. Ai-Bing Z, et al. An electron analyzer, http://144.206.159.178/ft/CONF/16421838/16421884. pdf. Accessed on Feb 23, 2013.
25. Lakdawalla E. China's Yinghuo-1 Mars Orbiter, http://www.planetary.org/blogs/emily-lakdawalla/2010/2655.html. Accessed on Mar 19, 2013.
26. ROSA: Radio Occultation Sounder Antenna for the Atmosphere, http://ieeexplore.ieee.org/xpl/login.jsp?tp=&arnumber=4458565&url=http%3A%2F%2Fieeexplore.ieee.org%2Fiel5%2F4446147%2F4458235%2F04458565. Accessed on Mar 17, 2013.

Part III
Depicting and Debating

Chapter 7
Mars Missions: Past, Present and Future

Is there life on Mars? No, not there either.

A famous joke form Soviet Era.

Since 1960s various missions to study Mars have been launched. Few recently launched missions are still active and are providing useful, relevant and unknown information about various aspects of Mars. Currently, few states are working on various new mission proposals for Mars, and many of such missions are expected to be launched between 2013 and 2020 periods. Few private organisations are found taking keen interest in various activities associated with different Mars programmes, and there exists a possibility that in future, even few private organisations could undertake missions to Mars. State agencies and private organisations are found exploring various innovative ideas to carry the Mars agenda further. This chapter offers a broad overview about some of such programmes and ideas. Here, attempt is not to provide every available detail about the past, present and future activities, but the effort is to offer a broad overview of various activities and discuss the likely trends for the future investments.

7.1 A Journey Through a Half Century

Space technologists started developing interest in visiting/studying Mars almost around the same time the space age began. Within three years after the launch of the first satellite Sputnik (1957), the first Mars mission was launched by the erstwhile USSR. It is important to appreciate the vision of mankind in respect of Mars has mostly been about humans visiting the Mars for many years. From literature to science fiction to movies to cartoon films, everywhere Mars has been always viewed as an "objective" to be conquered. Particularly, the success of Apollo programme to put humans on Moon made people believe that the Mars cannot be far-off. Most importantly, since 1950s onwards, some portion of the scientific community has been concentrating towards developing various technologies essential for human visits to Mars. However, presently the world is much

---

*Catling DC. Planetary science: on Earth, as it is on Mars? Nature 2004;429:707–708.

A. Lele, *Mission Mars*, SpringerBriefs in Applied Sciences and Technology, DOI: 10.1007/978-81-322-1521-9_7, © The Author(s) 2014

far-off from the human Martian travel to happen at least in near future. Technology and cost could be the two most important factors for this not to happen. During 1970s, the soaring costs of Vietnam War had impacted the development of the project JAG which was about the manned flyby mission to Mars. Subsequently, end of Cold War has significantly diminished the "notion of technology one-upmanship race". The argument of "economics" gained more relevance during last few decades and states started making "measured" investments in science and technology. In the past essentially, the policies and vision of the then United States President Kennedy were responsible for Apollo to happen and since then the world is in search of a Kennedy for Mars!

The erstwhile Soviet Union was first to develop a mission to study Mars. However, the first attempt by the Soviet Union at launching the interplanetary probes was a failure. During October 1960, they had launched Korabl 4 and 5 missions. But, the third stage of both launch vehicles had failed. Few subsequent attempts for flyby missions also failed. The first success came to NASA in 1964 when the Mariner 4 spacecraft took close-up pictures of Mars. It took pictures while passing within 9,844 km of Mars. The mission lasted for almost three years. Within next few years, NASA had a success with two more identical missions. Mariner 6 flew by Mars at an altitude of 3,431 km and Mariner 7 at 3,430 km. During 1971, Mariner 9 became the first aircraft to orbit around another planet. In fact, Mariner 8 and 9 were the third and final pair of Mars missions in NASA's Mariner series of the 1960s and early 1970s (Mariner 8 was a failure). Both were designed to be the first Mars orbiters, marking a transition in our exploration of the red planet from flying by the planet to spending time orbiting around it. This mission that lasted for one year but was not able to collect pictures initially for almost 45 days, because a huge Martian dust storm had engulfed Mars. This mission for the first time provided the close-up views of both the moons of Mars [1]. Mariner 9 exceeded all primary photographic requirements by photomapping 100 % of the planet's surface [2].

NASA's Viking Project was the first United States mission to land a spacecraft safely on the surface of Mars and return images of the surface. This craft was launched on Aug 20, 1975, and it landed on Mars on July 20, 1976. This project constituted of the launching of two identical spacecrafts (Viking 1 and 2), each consisting of a lander and an orbiter. This pair flew together and entered Mars orbit; the landers then separated and descended to the planet's surface and landed at different places. Apart from routine observations, they also collected vital science data from the Martian surface and conducted three biology experiments designed to look for possible signs of life. However, these experiments provided no clear evidence for the presence of living microorganisms in soil near the landing sites. This entire mission was to last for 90 days but successfully operated far beyond its design lifetime almost for four years [3].

Unfortunately, subsequent to Viking 1 and 2 success, both the erstwhile USSR and the United States had to encounter series of failures with their Mars missions. NASA's Mars Global Surveyor (1996/97) became the first successful mission to the red planet in two decades. It took a year and a half for this craft to trim its orbit from a looping ellipse to a circular track around the planet. Finally, the spacecraft could begin its prime mapping mission only by March 1999. The mission conducted a detailed study of the entire Martian surface, atmosphere and its interior. This mission played an important role towards increasing the knowledge about meteorological conditions on Mars. It has also provided the 3D views of Mars' north polar ice cap [4]. This mission operated longer at Mars than any other spacecraft in history and for more than four times as long as the prime mission originally planned and went silent only in November 2006 [5]. The spacecraft returned exhaustive information that has smartened up the understanding about Mars. The assessment provided based on the data collected also offers help to select landing sites for the robotic Mars missions.

In 2003, the ESA had received a partial success with its Mars Express Orbiter/ Beagle 2 Lander mission. This mission consisted of two parts: the Mars Express Orbiter and the Beagle 2 Lander. Unfortunately, the lander failed to land safely on the Martian surface. However, the orbiter has been successfully performing scientific measurements since early 2004. Nearly 90 % of Mars' surface has been mapped by the high-resolution stereo camera on Mars Express [6]. The mission is successfully completing 10 years of its existence. It is expected that in coming few years, remaining Mars surface would be mapped.

Apart from ESA's Mars Express and a failed attempt each by Japan and Russia/ China, the Mars has been the singular United States-dominated agenda post 1996. NASA has achieved success with various missions like Mars Odyssey (2001), Mars exploration missions like Spirit and Opportunity (2003) and reconnaissance orbiter (2005). Phoenix Mars Lander (2007) is another important mission which has provided significant amount of information. One of the spectacular achievements by the United States has been the launch of Curiosity mission (Mars Science Laboratory).

Curiosity mission was launched on 26 November 2011, and this Science Laboratory almost to the size of a big truck arrived on Mars on 5 Aug 2012. This entire mission could be viewed as a technological marvel both from the point of view of landing an object of such size and shape on the Mars surface by using precision landing techniques and also for the quality of the Science Laboratory (payload capability), which is currently performing a range of experiments on the Martian surface. The mission also has a truly international flavour, with various instruments onboard belonging to states like Russia, Canada, and Spain. This is the first planetary mission to use precision landing techniques, steering itself towards the Martian surface similar to the way the space shuttle controls its entry through the Earth's upper atmosphere. For the landing of the Curiosity rover on Mars surface in the final minutes before touchdown, the spacecraft activated its parachute and retro rockets before lowering the rover package to the surface on a tether [7]. Presently, Curiosity is leading very busy life over Mars and collecting and analysing information. Some early results based on analysis of a rock sample collected indicate that ancient Mars could have supported living microbes.

It is expected that in the coming decade, few more Mars missions could be launched and planning for most of these missions has already began.

7.2 Future Programmes

7.2.1 Maven

In order to support innovation, NASA had formulated the Mars Scout programme to send a series of small, low-cost robotic missions to Mars. As a first mission under this programme, Phoenix lander was launched on Aug 4, 2007. The proposed second mission is MAVEN (Mars Atmospheric and Volatile Evolution) scheduled for a launch window during 18 November–7 December 2013 and Mars orbit insertion around 16 September 2014. The neutral gas and ion mass spectrometer (NGIMS) instrument that will measure the composition of Mars' upper atmosphere has already been integrated into the MAVEN during the first week of April 2013. NGIMS will measure the composition of neutral and ionised gases in the upper atmosphere as the spacecraft passes through it on each orbit and will determine the basic properties of the upper atmosphere. NGIMS can improve the understanding of the history of Martian climate, if conditions on early Mars may have been conducive for supporting microbial life [8].

The goal of MAVEN is to determine the history of the loss of atmospheric gases to space through time, providing answers about Mars' climate evolution. By measuring the current rate of escape to space and gathering enough information about the relevant processes, scientists will be able to infer how the planet's atmosphere evolved [9]. This mission programmed to study current rate of atmospheric loss was originally supposed to be a "double" in which two orbiter units were to be sent but NASA is currently only planning to fund one such mission. MAVEN would be the first spacecraft ever to make direct measurements of the Martian atmosphere [10]. One of the key sensors in this mission is the MAVEN magnetometer, which would play a key role in studying the planet's atmosphere and interactions with solar wind, helping answer the question of why a planet once thought to have an abundance of liquid water became a frozen desert. This magnetometer would be the key to unravelling the nature of the interactions between the solar wind and the planet [11]. Another NASA lander mission planned for launch in 2016 is known as In Sight, Interior Exploration using Seismic Investigations, Geodesy and Heat Transport. The purpose of this mission is to study early geological evolution of Mars.

In order to generate the interest of particularly the younger generation towards ongoing Mars exploration in general and MAVEN mission in particular, NASA has invited members of the public to submit their names and a personal message online for a DVD to be carried aboard a spacecraft that will study the Martian upper atmosphere. The DVD will be in NASA's MAVEN spacecraft [12].

7.2.2 ExoMars (Exobiology on Mars)

This is one of the important and most debated missions for near future. This mission was originally approved by ESA in 2005 but has gone through multiple phases of planning with various proposals over the years. On 14 March 2013, the ESA and Roscosmos (Russian Federal Space Agency) have formally signed an agreement that will take the two space agencies to Mars on two interconnected missions: one in 2016 (an Orbiter mission) and the second during 2018 (Rover mission). The basic purpose of both these missions is to gain knowledge about Martian environments and search for possible biosignatures of life on Mars. These missions have generated good amount of interest among the ESA member states, and even the new ESA members like Poland and Romania are expected to contribute up to €70 million for this programme [13]. NASA was expected to be major contributor in this programme; however, under the financial year 2013 budget released by President Obama on 13 February 2012, NASA has terminated its participation in ExoMars due to budgetary cuts. Now NASA is expected to do a nominal contribution and that too for the 2018 mission.

The 2016 mission constitutes of the delivery of a stationary lander and orbiting satellite to Mars. The satellite, the Trace Gas Orbiter (TGO), would scan the Martian surface for sources of Methane and other gases indicative of biological life of the kind we have on Earth. The assessment made by TGO would help narrow down the choice of landing site for a Martian rover to be launched in the second part of the mission in 2018. The stationary lander (EDM) will have a host of instruments from both ESA and RosCosmos on it, designed at helping the second rover in 2018. The anticipated science package on the lander is called DREAMS—Dust characterisation, Risk assessment and Environment Analyser on the Martian Surface. DREAMS will analyse the Martian atmosphere in terms of humidity, temperature, wind velocity and dust content while also carrying a camera to check the effect of Martian dust on visibility. The final sensor is quite interesting, measuring the effect of electrical charge on dust in the atmosphere (ARES).

In 2018, the actual ExoMars robotic rover will be launched atop a Russian Proton rocket. The rover itself is packed with instruments designed to find life up close and below the surface of Mars, where it would be shielded from the red planet's harsh external environment. Onboard, the rover would have imaging systems like panoramic camera, infrared spectrometer and remote sensing and drilling mechanisms [14]. Overall, the ESA is expected to provide the probes and RosCosmos, the launching rockets and descent mechanisms to deliver the probes to the surface of Mars. NASA could provide the some support in regard to the communications equipments.

On the whole, it is expected that apart from the United States, few other states like Russia, Japan, China, and India could launch their own Mars missions in future and would seek to exploit the opportunities in 2016 and 2018. The list of various past, current and future Mars missions is attached as Appendix A.

7.3 Human Mission

The general direction of the debate on deep-space issues brings out likelihood of two plausible futures. First, around 2030: China could be attempting to undertake the human Moon mission. Second, the United States may attempt a human Mars mission around the same time. However, these are only predictions and various issues form political will to technology development to budget provisions would play a vital role for them to become a reality. Presently, some scientific groups, state space agencies and private enterprises are found designing and developing ideas for human deep-space mission. In fact, this activity is happening since 1960. It is important to take a note of such activities because in future some of such ideas are only expected to emerge as possible options.

The inspiration of Mars flyby using 500-day orbit was conceived during 1962 by various studies by NASA and outside aerospace contractors under the project Empire. During 1966, a plan for 1976 mission called JAG was proposed for a nuclear-powered rocket carrying four humans on a flyby around Mars. This was a 667-day voyage where one part of the mission was involved dropping an auto-mated instrument on the planet to collect soil samples. During 1960-70s NASA had developed a significant reputation for itself with its Apollo success and their proposals were taken seriously by everyone. Subsequently, they had mooted an idea of winged shuttle, a space station and a large human expedition to Mars. However, the entire plan could not become a reality with President Richard Nixon approving only the development of the space shuttle. The Bush administration (1989) was asked for a support for a Mars expedition materialising by 2010. But, the cost estimates soared to over $500 billion, dooming this effort. The proposal of developing Moon–Mars programme with human trips to moon by 2015 and to Mars around 2030 was not accepted during 2004 by the United States administra-tion again due to budget constraints [15]. Probably, cost and lack of political support are the key factors for the human Mars mission not to happen till date. Post "Curiosity" success, it appears that both state and commercial players could get motivated to invest in human mission to Mars. Also, there exists a possibility that private–public partnership could push the Mars agenda further.

The United States aerospace giant, the Boeing Corporation, has outlined their view of what technologies could be used to accomplish humankind's goal of vis-iting crews to the Martian system. They want to develop a connection between the lunar missions and the proposed Mars missions through a phased exploration. They are also suggesting few "new technologies and innovations", which includes concepts like inflatable Crew Transfer Habitation (CTH) module [16]. It is important to note that NASA supports and offers contracts to the business house like Boeing and others, and eventually, the commercial judgment of Earth–Moon–Mars economy may lead the future of human missions to Mars.

The billionaire founder of the private space company (Space X), Elon Musk, has also disclosed his plans to build a colony for 80,000 people on Mars. He proposes to send people for one way to Mars for $500,000 each. Under the project

Red Dragon, an automated "Dragon" vehicle is expected to land on Mars in 2018, paving the way for an eventual human landing. Also, the first space tourist Dennis Tito is keen to undertake a 501-day round trip to Mars. His Inspiration Mars Foundation, an American non-profit organisation, aims to launch a flyby human Mars mission on 5 January 2018. It has been also reported that the National Geographic Society is exploring the idea of associating with Tito's group. There is possibility that a man and a woman would be the first two professional crew members. It would be a "fast, free-return" mission, passing within 100 miles of Mars before swinging back and safely returning to Earth.

Apart from designing and planning technologies for futuristic humans' Mars visits, some interesting work is being undertaken to simulate the Martin atmosphere on the earth. In the deserts of Utah, in the Western United States, a project called the Mars Desert Research Station (MDRS), a simulated off-world habitat that serves as a test site for field operations in preparation for future human missions to Mars, is under progress. The "astronauts" are a group of volunteers who are helping to discover ways to investigate the feasibility of a human exploration of Mars and use the Utah desert's Mars like terrain to simulate working conditions on the red planet [17]. Generally, it has been observed that the issue of one-way human mission to Mars is being discussed [18] as one of the most viable options under the existing technological capabilities by few individuals and agencies. It appears that many volunteers would be available if such idea becomes a reality in near future.

China is yet to announce any major plan about its Mars vision. However, as mentioned in one of the earlier chapters their scientific community has been in performing few interesting experiments in respect of future colonisation of Mars. They are experimenting with the idea to grow fresh vegetables on the moon and Mars. Chinese researchers have successfully completed the preliminary testing in this regard in Beijing. They have completed the initial testing of an "ecological life support system", a 300-m^3 cabin providing the atmosphere akin to Mars [19]. The system was able to grow four types of vegetables successfully. Few other countries have tested such systems in the past. China's long-term interest in Mars also becomes evident from its participation in the Mars-500 experiment (2007–2011). In this experiment, a Chinese person had participated and had stayed in a simulated Martin atmosphere for 520 days to check the feasibility of human endurance. All this indicates that China is attempting to "test the water" for the present and may eventually develop a comprehensive Mars programme.

7.4   Outlook

For all these years, the success rate for Mars missions has not been very encouraging, but during last few years, the situation has improved. Also, there have been few partially successful missions. Presently, budget appears to be the main constraint for various states. Human mission has been a subject under greater discussion for many decades, but actually no major effort has taken place in this regard. Probably, by end of current decade, there exists a possibility that at least a flyby human mission may become a reality. With new states and particularity the private players entering into the "Martian atmosphere", it is expected that interesting days are ahead for Mars research.

References

1. http://www.planetary.org/explore/space-topics/space-missions/missions-to-mars.html. Accessed Jan 2, 2013.
2. http://mars.jpl.nasa.gov/programmissions/missions/past/mariner89/. Accessed Dec 9, 2012.
3. http://nssdc.gsfc.nasa.gov/planetary/viking.html. Accessed Dec 9, 2012.
4. http://mars.jpl.nasa.gov/MPF/. Accessed Dec 9, 2012.
5. http://www.nasa.gov/mission_pages/mgs/mgs-20070413.html. Accessed Feb 12, 2013.
6. Mapping Mars. Feb 06, 2013. http://www.marsdaily.com/reports/Mapping_Mars_999.html. Accessed Feb 26, 2013.
7. Mars Science Laboratory (MSL) Spacecraft. (http://mars.jpl.nasa.gov/msl/). Dec 9, 2012.
8. Nancy Neal Jones. Final MAVEN instrument integrated to spacecraft. Apr 5, 2013. http://www.marsdaily.com/reports/Final_MAVEN_Instrument_Integrated_to_Spacecraft_999.html . Accessed Apr 6, 2013.
9. http://www.nasa.gov/mission_pages/maven/news/magnetometer.html. Accessed Apr 14, 2013.
10. http://www.nasa.gov/home/hqnews/2007/dec/HQ_07283_Mars_Scout_Missions.html and http://mars.jpl.nasa.gov/programmissions/missions/future/maven/. Accessed Mar 1, 2013.
11. http://www.nasa.gov/mission_pages/maven/news/magnetometer.html. Accessed Apr 14, 2013.
12. NASA Invites Public to Send Names and Messages to Mars. May 03, 2013. http://www.marsdaily.com/reports/ NASA_Invites_Public_to_Send_Names_And_Messages_to_Mars_999.html. Accessed May 4, 2013.
13. http://www.spacenews.com/article/europe-pinning-its-exomars-hopes-patchwork-funding-sources#.UVu52HbrbIU. Accessed March 10, 2013.
14. http://astronomyaggregator.com/spaceflight/esa-and-roscosmos-formalise-mars-exploration-partnership/. Accessed Apr 2, 2013.
15. Bill Douthitt, http://news.nationalgeographic.com/news/2013/02/130228-manned-mission-mars-psychology-space-science/. Accessed Mar 6, 2013.
16. Gebhardt C. Boeing outlines technology for crewed Mars missions. Jan 18, 2013. http://www.nasaspaceflight.com/2013/01/boeing-outlines-technology-crewed-mars-missions/. Accessed Mar 10, 2013.
17. Sam Webb. The humans who are exploring 'Mars' already—researchers in space suits recreate red planet base in the Utah desert. 12 March 2013. http://www.dailymail.co.uk/sciencetech/article-2291725/Humans-exploring-Mars-Amazing-pictures-mission-simulation-base-UTAH.html#ixzz2NJS90ln5. Accessed Mar 12, 2013.
18. Davies P, Schulze-Makuch D, editors. A one way mission to Mars colonizing the red planet. Cambridge: Cosmology Science Publishers; 2011.
19. Prigg M. Marrows on Mars: China reveals plans to grow vegetables in extra-terrestrial bases. Dec 4, 2012. http://www.dailymail.co.uk/sciencetech/article-2242792/Marrows-Mars-China-reveals-plans-grow-vegetables-extra-terrestrial-bases.html. Accessed Jan 2, 2013.

Chapter 8
Cost Contested: Perceptions Versus Reality

> *The cost of London (2012) and Beijing (2008) Olympic was*
> *approximately USD$15 billion each and the total cost of*
> *Nasa's Mars Science Laboratory, of which the Mars Curiosity*
> *Mission is a part of, is about: US$2.5 billion.*

The correct utilisation of the tax payer's money has always been the most passionately debated subject internationally. Particularly, during various deliberates on issues concerning the investment in space technologies this concern always gets reflected equally for the developed states like the United States and also for the developing states like India. Such debates become more intense when the investments are seen made in so-called glamorous projects like building a space station or humans staying in space or conduct of deep-space missions, particularly the missions like the Moon and Mars missions. The basic concern always has been the issues of economic viability and rational for undertaking such activities. Various "for and against" arguments and counterarguments have been made by many in support of their position. However, all such investigations are mostly found being made based on certain perceptions. What is important is to delink such a scrutiny from emotions and carry out a more pragmatic assessment. This chapter debates the issue in regard to the efficacy of monetary investments in space science in general and Mars in particular.

In the Cold War era, space exploration was part of a proxy competition for geopolitical rivalries. Both during the Cold War period and in the post-Cold War period, space exploration programmes have received criticisms for various reasons ranging from cost to safety of the individuals involved. However, interestingly there are many advocates of space mission too. More importantly, the public opinion in many countries has been usually supportive of these programmes; however, at times it has been observed that some NGOs or individuals with mostly lopsided ideas about socialism and development criticise such investments and media give them unneeded publicity.

Space missions have resulted in a variety of important discoveries, including the effects of low gravity on humans, the presence of Van Allen belts around the

---

*http://www.captaincynic.com/thread/94243/cost-comparison-mars-curiosity-vs-olympics.htm, accessed on 16 April 2013.

A. Lele, *Mission Mars*, SpringerBriefs in Applied Sciences and Technology, DOI: 10.1007/978-81-322-1521-9_8, © The Author(s) 2014

earth, images of the far side of the Moon and the absence of intelligent life on Mars [1]. Needless to say that everybody understands the revolutions made by the satellite technologies in the field of communications, remote sensing and navigation. Presently, a wider debate is taking place in vibrant democracies like the United States and India in respect of need to invest money for undertaking "exotic" space missions like mission to Mars when the same money could be utilised for addressing urgent needs of the mankind on the earth. Current discussions also revolve around the possibility of future space colonisation—that is, the establishment of human settlements on Moon or Mars.

Few important arguments those are advanced in any discussion about the utility of space exploration and the roles of humans and robots could be summarised as follows [2]:

1. Space exploration would eventually allow mankind to establish a human civilisation in another world (e.g. Mars).
2. Various states explore space and create important new technologies to advance the economy. The money spent on space programmes has given much in return both in economic terms and otherwise. Space exploration also serves as a stimulus to attract young minds to enter the fields of science and engineering.
3. Space exploration in an international context offers a peaceful cooperative venue that is a valuable alternative to nation state hostilities. The best showcase for this idea is the International Space Station, an experiment where the Cold War era rivalry has been converted into a multilateral collaborative effort where the erstwhile Soviet Union and the United States are active partners.
4. Space programme contributes significantly towards increasing the national prestige. Historically, the great civilisations have never dared to abandon exploration and Mars missions are the continuance of this policy.
5. Exploration of space could provide humanity with an answer to the most fundamental questions: Are we alone? Are there other forms of life beside those on Earth?

In the present era, it is important to appreciate the fact that people may not probably believe the efficacy of a Mars mission entirely based on science agenda alone. Would this be happening because of the limited appetite of the average population for the developments in the science and technology? The answer is probably not. Maybe there are few disbelievers for various reasons. However, what is found lacking is the overall understanding of the issue and that maybe happening because the correct message and information has not reached to the people (critics) in time. Hence, there is a need to find the right balance of science and technology and other associated factors to offer the people logical and correct information.

Various disciplines of knowledge could be exploited to unearth a rational for the states making investments into the space sciences. There could a need to develop a broad philosophy-based approach to argue the case, and an attempt needs to be made to get the message that, "Exploration is intrinsic to our nature. It is the contest between humans and nature mixed with the primal desire to conquer. It fuels curiosity, inspiration and creativity. The human spirit seeks to discover the unknown, and in the process explore the physical and psychological potential of human endurance". For all these years, it has been observed that, "There have always been the few risk-takers who ventured for the rest of us to follow. Because of earlier pioneers, air travel is now commonplace and space travel for all is just around the corner" [2]. Essentially by opting to visit Mars, space scientists are not changing any course of research but are essentially raising the bar in regard to their ongoing work. The exploration of space over the years has been a boon for the mankind, and everyone agrees that the human have got in return much more that what they have invested. Essentially, the same logic needs to be extrapolated for the Mars agenda too.

Interestingly, the critics to Mars mission are found coming from different walks of life. It has not been always that only people with less knowledge of science or from an economically backward background or people mainly with skewed developmental ideas or people from non-governmental organisations (NGOs) with a "different" worldview are the main critics of such missions. Few people from different walks of life with a respectable financial and educational background and belonging to the so-called affluent layer of society are also found criticising the relevance of such missions, essentially giving an indication that they are less informed on the subject.

One of the celebrity that has recently disapproved the NASA's Mars mission agenda is the Austrian daredevil Felix Baumgartner (known as fearless Felix)who jumped from 24 miles above earth, breaking the speed of sound (he reached the speed of Mach 1.25, or 1.25 times the speed of sound) before releasing his parachute on 14 October 2012. Felix who was supersonic for a half-minute has surprisingly emerged as an ardent critic of Mars missions. He believes NASA's Mars programme is a waste of money that money could be used for "saving our planet". NASA landed the Curiosity rover on Mars in 2012 to carry out a $2.5 billion exploration mission over the next 10 years.

Mr. Felix Baumgartner has a right to express his opinion, and his viewed should be respected. However, it is also important to reveal that his one event lasting for minuscule of time called Red Bull Stratos probably coasted $30 million, according to one estimate (officially cost of the event has not been made public)[1] [3]. One way to look at his mission is that a significant amount of money was sped only to set some obtuse world record which is actually a trivial activity. But conversely, it could be argued that such acts only assure about human determination and endurance and they are the real motivators for the society in general and youths in particular. Interestingly, this entire experiment could offer some knowledge to help future space crews in regard to surviving any unfortunate high-altitude accidents.

Undoubtedly, Mr. Felix Baumgartner is a hero, but his assessment about the financial costs of Mars mission is erroneous. This particular case has been mentioned over here not from the point of view of criticising and isolating an

---

[1] It has been reported that the cost of the helium for balloon filling was $70,000 and cost of each balloon was $250,000. His space suit coasted $200,000.

individual but to demonstrate a fact that how people directly associated with the issues related to aeronautics and aerospace engineering are also less informed about the relevance of other activities taking place in the field of their interest. This essentially indicates that for the development of certain (incorrect) perceptions in regard to the Mars mission, the scientific community could also be equally responsible for not spreading the correct message across in time.

Today, humans are trying to learn more about the science of Mars not just to gain additional knowledge about planet Mars, but such knowledge could assist to know more about the Earth too. Mars has undergone radical transformations because of climate change, volcanoes, asteroid strikes and various other events. Earth has also witnessed a very turbulent past, and we are keen to know what would happen in future. Mars could provide us some clues in this regard. Hence, we have much to learn from the evolution and presence of Mars in order to secure the future of the earth.

The manned Moon mission during 1960s has made tremendous impact on the development of science and technology not only for the United States but globally. It was one of the most challenging missions then (even today reaching Moon is a major technological challenge), but scientific community in general and society in particular grow only when they are challenged. In 1960s, the scientific community in the United States was challenged by then President Kennedy and Apollo programme is a result of that. Mars offers the paramount challenge at least for now. Investing in Mars is also investing in future generations. Investments into science show a dedication to intellect and experimentation. Providing the resources to dig into the universe ultimately assists to make life on earth better. Such investments push humans forward, triggers inventions, medical breakthroughs, environmental discoveries, green technologies and so much more than just Martian soil samples [4]. Most of such benefits are long-term benefits, but it needs to be realised that it is important to invest today then only we could think of long-term benefits. The critics for missions like Mars are probably thinking very narrow and want some immediate visible tangibles. Hence, without realising the "value" of such missions, they are putting a "cost" towards it.

It is also important for the scientific community to justify their investments by providing a cultural argument too simply because it could have a far better appeal for the masses than any economic or scientific justification. This is because culture is the way of life and depicts conventional patterns of thought and behaviour, including values, beliefs, rules of conduct, political organisation, economic activity and the like, which are passed on from one generation to the next by learning and not by biological inheritance and could assist the humans to appreciate the larger dynamics of importance of space science [5]. For more than five decades, humans are investing in the space area and have achieved multiple successes. The progress made in space sciences has benefited humanity tremendously and played a major role towards the overall sustainable development. The growth of technology had also helped to bring in a new culture of rationality and helped to overcome various superstations. The success of Apollo has helped to institute a new culture of studying planetary science not just from the perspective of astrophysics but to analysis the geophysical structures of planets and experimenting in space from a point of view of establishing human colonies on planets in future.

In the nineteenth century, few scientists referred the term "culture" to refer to a universal human capacity [6]. It is important to appreciate that culture plays a vital role in making progress and increasing creativity and must be carefully nurtured to grow and develop. Public expenditure on space science can and should be justified largely on cultural grounds. The globalisation of space science helps, and it is relatively easy to convince most people that mankind as a whole should continue to explore this frontier of knowledge and can afford to do so. It is important to appreciate that the masses generally find the cultural argument more convincing than any other arguments [7]. Human interest in outer space is not new. Since historical times, humans have always shown interest to know more about the universe, and this interest over a period of time has only been beneficial to the mankind. Hence, investments in Mars mission are noting but the continuation of the historical quest of mankind to know more about the universe.

While justifying the global investments into Mars mission, it is also important to question oneself that "are the arguments based on the suggestion that any further research in science and technology would always be beneficial to the mankind" always true? Is the assumption that such investments are necessary to make the future of humanity superior is actually doing injustice to the present? It could possible to present an argument both ways. However, what is important is to make an overall assessment by considering various factors and not to have any stringent positions. It is difficult to carry out a typical cost-benefit analysis because the effort is to know more about various unknowns. Any cost-benefit assessment carried out from only the economic perspective to be confusing.

One essential question needs to be answered and that it, "is there a danger of 'progress trap'[2] over here"? A progress trap is the condition the human societies experience when, in pursuing progress through human ingenuity, they inadvertently introduce problems they do not have the resources or political will to solve, for fear of short-term losses in status, stability or quality of life. This prevents further progress and sometimes leads to collapse. It is argued that in a progress trap, those in positions of authority are unwilling to make changes necessary for future survival. To do so, they would need to sacrifice their current status and political power at the top of a hierarchy. They may also be unable to raise public support and the necessary economic resources, even if they try. Mars missions are futuristic in approach. Naturally, the scientific community and the decision makers involved in this activity today may not be at the helm of affairs when the actual benefits from these investments would become evident. Most importantly, technocrats of the day are arguing for Mars belong to the category of people in the

---

[2] Ronald Wright in 2004 gave a series of lectures in the Massey series on this issue, and he is famous for his work titled *A Short History of Progress* on the same subject.

favour to make changes necessary for future survival. Hence, supporting Mars mission is not about falling in any so-called progress trap.

As humans, we value progress and believe in moving forward, no matter what the cost could be. At times the technological innovations helping the society could also bring in new problems which are difficult to solve. Like certain technological developments end up adding to environmental problems or technologies like laser. having their own drawback. Missions like Mars mission are unlikely to offer any technological side effects, and also Mars being just one step towards the lager goal of discovering the universe, it is unlikely that humans would fell victim to their own success with so much more left to achieve.

Investing in Mars is not only learning about Mars. Such investments could offer an access to range of other applications. The technologies developed for various Mars missions in the past, and the ongoing research and development for future Mars missions have certain direct or indirect utility for improving life on earth and also for the usage in other space missions. Life-saving robots, panoramic cameras and pathogen-detecting sensors are just a few of the remarkable spin-offs to emerge from these efforts, with many more sure to come before the first human sets foot on Martian soil [8]. Essentially, the sensor technology and the robotic technology are the two main areas of technology development where whatever being developed for Mars could find utility in other parts of life. Establishing communications with the missions to Moon and Mars is one of the foremost challenges. States investing in such missions have developed deep-space networks for the communication. All this could allow them to develop new type of high-level information services. Also, overall increase in both hardware-and-software-application-related knowledge in regard to data networking and data processing would increase.

However, it is also important not to "stretch" the spin-off argument unnecessarily. Spin-offs could offer side benefits, but there are probably more cost-effective ways to get all of these spin-offs without leaving earth. Once Carl Sagan had argued that "you don't need to go to Mars to cure cancer" [2]. What is important to emphasise is that we are exploring space (Mars) to explore space (Mars). Spin-off is secondary argument and definitely has its own merit. However, spin-offs are incidental and should be the core argument. The key focus of various anti-Mars mission arguments is the case of economic viability of the mission and that the money is being unnecessarily diverted to such mission which is actually required for various social needs.

The basic question is, "does the benefit outweigh the costs?" There is no straightforward answer to such a quarry. There is a need to appreciate that the economic and societal benefits are unlikely to be immediately evident, but they would always follow [2]. When we invest in science, when we invest in space, when we invest in exploration, we always get far more back in return than we put in and not just in dollars and rupees but in various other ways [9]. Apollo programme unquestionably created unintended economic consequences and benefits, such as the by-product of miniaturisation, which led to developments in computers and cell phones. Apollo also created development of new infrastructure, subsequently offering increase in economic activity [2]. Overall, if various programmes are managed well, then they have the capability to offer multiple benefits in surplus to the original invested money.

In case of democratic dispensations, the key question which needs to be asked: "Is the government responsive to public opinion of space policy?" Broadly, the assessments in regard to this issue carried out since 1995 for specific case of NASA indicates that "the public supports the idea of space exploration, while also feeling that spending on space exploration is "too high". Therefore, the government appears to be giving the people exactly what they want in regard to NASA's budget—more money each year—but at the same time a smaller percentage of the federal budget" [10]. Probably, other states could also have a similar approach.

Indian space programme when began in mid-1960s came across a severe criticism by a certain section of people both from within and outside the country. The argument was simple that for a poor country like India, there is no need to join the elite group of space-faring group of nations. Still, India went ahead with its programme with a clear vision in mind that why they are making such investments in spite of all odds. Its initial investments never offered them any direct economical benefits. But the social benefits were tremendous from meteorology, to communication, to education. All this eventually played a role in the economic development of the state. During last four decades, India has established itself as formidable space power. All this became possible because its forefathers had made correct and timely investments with a vision. Today, India is being again criticised by few for its Mars agenda. Maybe in some way, the criticism in the 1960 and 1970s could be justifiable because it was an era when humans were actually not aware about the potential of the space technology and neither they had experienced it. Also, India's economic situation was appalling then. But the criticism in the twenty-first century is bit difficult to understand.

Over the years, India has succeed in raising its stature both economically and otherwise. Presently, Indian is regarded as one of the major and stable economies in the world. In space arena, ISRO has gained significant reputation. India's approach to space has been pragmatic and austere [11]. Since beginning, the country has pursued a space programme as a part of its socio-economic development model. ISRO has major focus towards developing remote sensing satellites and communication satellites. Today, ISRO is a world leader in the arena of remote sensing. India started its space journey with very limited research facilities and financial support from the government. However, today India has established itself significantly both in the context of hardware and in the context of software and has well-trained human capital available. India's first Moon mission was one of the most effective and cost-effective missions in the world. Under present settings, it could be even argued that today, India is one of the most eligible countries (technologically and economically) in the world to experiment in deep space and undertake missions like Mars orbiter.

The Indian Prime Minister Dr Manmohan Singh has justified the investments made into the country's space exploration, research and development programmes and citing technology prowess as a sine qua non for economic growth and inclusive development. He had articulated these thoughts after the announcement of India's Mars mission. As per him, the status of development depends on the technological abilities of the country, and in case of India with its proven technological competence, ISRO has played an important role towards India's progress [12]. The present Indian Mars agenda is devised more for learning: "how to undertake such challenging missions". Hence, the mission is also been called as a technology demonstrator. It is important to have such foundations for new discoveries and new knowledge. This entire process of basic research would build up a base for the future developments. All this would not only make the nation wiser about the Mars but would also bring in benefits both for social and industrial sector.

India's Mars mission is likely to cost approximately Rs 400 cores (approx. US$ 80–85 Mn). In fact, the same amount was also spent on India's first Moon mission in 2008. The total expenditure on India's deep-space programme has been around 2–5 % of ISROs overall budget [13]. The cost of Mars mission is expected to be 0.01 % of the country's overall annual budget [14]. In fact, with such a limited budget investments, India's mission to Mars is opening the new frontiers of space in India and the country should make full use of its expertise and evolve further.

References

1. http://www.newworldencyclopedia.org/entry/Space_exploration. Accessed on Jan 10, 2013.
2. Can space research be justified nowadays? 13 Nov 2011, http://cjcpig.wordpress.com/2011/11/13/can-space-research-be-justified-nowadays/. Accessed on Dec 14, 2013.
3. http://www.theaustralian.com.au/news/world/felix-baumgartners-plunge-from-stratosphere-breaks-broadcast-records/story-fnb64oi6-122649691367. Accessed on Mar 11, 2013.
4. Dockter J. An open letter to Felix Baumgartner, http://www.huffingtonpost.com/jake-dockter/felix-baumgartner-mars_b_2056635.html. Accessed on Mar 10, 2013.
5. Young RM. Mental space. London: Process Press Ltd; 1994.
6. http://en.wikipedia.org/wiki/Culture. Accessed on Mar 20, 2013.
7. Smith CHL. The use of basic science: benefits of basic science, http://public.web.cern.ch/public/en/about/BasicScience3-en.html. Accessed on Jan 23, 2013.
8. http://spinoff.nasa.gov/pdf/Mar_web.pdf. Accessed on Mar 20, 2013.
9. Plait P. This is why we invest in science, 21 Mar 2012, http://blogs.discovermagazine.com/badastronomy/2012/03/21/this-is-why-we-invest-in-science-this/. Accessed on Dec 12, 2013.
10. Steinberg A. Space policy responsiveness: the relationship between public opinion and NASA funding. Space Policy. 2011;27:240.
11. Giri C. Gaining from space, http://www.gatewayhouse.in/publication/gateway-house/features/gaining-space. Accessed on Feb 10, 2013.
12. PM justifies investments in space science, Deccan Herald, Sep 9, 2012.
13. Ramachandran S. India sets its sights on Mars, 19 Apr 2007, http://atimes.com/atimes/South_Asia/ID19Df02.html. Accessed on Jan 10, 2013.
14. Laxman S. Mars beckons India. New Delhi: Vigyan Prasar; 2013. p. 5.

Chapter 9
Wrapping Up

> Space-faring nations have the responsibility to coordinate their
> efforts to launch time-bound, financially shared programmes to
> take up societal missions on a large scale, pooling their
> capabilities in launch vehicles, spacecrafts and applications.
> Such major cooperation will act as a great measure of security,
> in addition to empowering the most underprivileged,
> minimising communication gaps and reducing threats of
> conflict.
>
> A. P. J. Abdul Kalam

For the mankind, the journey in space has been fascinating for last five decades. Particularly, few states have achieved significant success in the short period of time in this field. The journey that had started with a launch of a small satellite into the low earth orbit (1957, Sputnik launch) subsequently peaked to the man reaching the Moon and now humans have even ambitions of capturing asteroid and changing the track of its motion! Today, states are using space systems to undertake various kinds of activities from undertaking various business transactions to disaster management to fighting wars.

Unfolding the mysteries of the Mars has always been a dream for humans since the beginning of space era. Since early 1960s, few states are found attempting with great vigour the Mars dream and have achieved some limited success. In recent times, few additional players (both states and private agencies) are found taking keen interest to invest in Mars missions.

The desire for knowing more about Mars could be the basic reason for human exploration of Mars. Broadly, it could be for the following reasons. First, at present, the technological expertise exists to attempt such missions and that has made states/non-states confident to invest more in Martian research. Second, the overall of success rate of Mars missions since 1960s has been around 50 % however, various recent successes indicate that the states have learnt from previous failures. Particularly, the success of the United States mission "Curiosity" has raised the expectations further. Also, now significant amount of information about Mars is available. This information is further being found helpful to plan future missions more appropriately. Third, apart from the USA and Russia, Asian states like China and India are found showing interest in learning more about Mars

---

*He is a former President of India and an eminent scientist, quoted in Lele A, Singh G (eds.)
Space security and global cooperation. New Delhi: Academic Foundation; 2009. p. 21.

A. Lele, *Mission Mars*, SpringerBriefs in Applied Sciences and Technology,
DOI: 10.1007/978-81-322-1521-9_9, © The Author(s) 2014

mainly because presently they are much better placed both technologically and economically to undertake such missions. Fourth, during last few decades, particularly China has significantly raised the bar for itself in undertaking various technologically challenging space missions. May be this is leading few other states to reorganise their space policies in general and deep space policies in particular. Fifth, private investors are also keen to peruse Mars agenda. This has raised the chances of establishing public–private partnerships on various projects in deep space arena. Six, humans view Mars as the most appropriate planet for detection of extraterrestrial life. Seventh, humans are keen to chase the dream of human settlement over Mars. Eighth, with more and more Earth-like alien planets (with chances of life) being discovered around the galaxy, it is high time for the humans to at least make their presence felt in their own solar system.

Mars has been most discussed and debated planet over the years however, mainly because of the technological challenges till date very few states have become successful towards developing capabilities to reach Mars and to undertake robotic landings. Space-faring states having relatively rudimentary capabilities to launch satellites are not in a position to undertake missions in deep space. A greater amount of expertise is essential to undertake complicated missions like Mars mission. Naturally, every space-faring state has not been found interested in attempting to reach Mars.

India became a space-faring state during 1980s. In spite of nascent beginning, during last few decades, Indian space programme has witnessed a remarkable growth. Few years back, India has successfully made forays into the deep space arena by undertaking its first Moon mission. Now, India has desires of Mars exploration and their first mission to Mars is waiting for a November 2013 launch. It appears that India is well prepared to take their first baby-step in the Martian orbit.

Generally, various space programmes have a longer gestation period. Hence, such programmes are planned much in advance and people usually have knowledge about the state's future space agenda. In case of India's Mars mission, there is a perception among the few that this mission has been hurriedly planed. Probably, there is a valid reason to "formulate" such opinion because there was not much debate about this mission prior to its announced on Aug 15, 2012. There could be two main reasons for this: one, the success of India's first Moon mission and particularity the finding of the presence of water on the Moon's surface had attracted much attention so much so that India's Mars agenda never came to the forefront, and two, officially also there was not much of talk on this subject.

India's second Moon mission (partnering Russia) was slotted for 2012 launch, and the main focus of ISRO was towards this mission. Because of some troubles at the Russian end and also due to the issues associated with non-availability of operational Geosynchronous Satellite Launch Vehicle (GSLV) from Indian side for the launch of this mission, presently this mission stands delayed and may take off by 2014. For last few years in Indian context, the focus of debate on space issues was mainly concentrated on the prospects of second Moon mission, ISRO's failures in regard to development of cryogenic engine technology and few alleged administrative failures in respect of some spectrum allocations by ISRO couple of years back.

Because of this what went unnoticed that in spite of few failures and controversies for other various ISRO laboratories, it was the "business was as usual". A dedicated team of ISRO was working on Mars mission without any fanfare. Interestingly, since 2007/2008, various media interviews by successive ISRO chiefs have mentions about India's interests in Mars mission. For long ISRO had identified and announced three launch windows for Mars mission—one each in 2013, 2016 and 2018 and it was obvious that ISRO would make its own choice. The March 2012 budget announcement by the government of India allocating Rs. 125 cores for the Mars mission made it obvious that India is serious about the Mars business and the mission could take place on the first available opportunity.

For the 2013 mission, India has already made it clear the main objective of mission is to establish Indian technological capability to reach Martian orbit. Further, during the orbital life of the spacecraft around Mars, scientific studies would be undertaken using scientific instruments on board the spacecraft on surface features, morphology, topology and mineralogy of Mars and the constituents of Martian atmosphere.

India's overall planning indicates that India is avoiding getting overambitious in the deep space arena but at the same time has done some sharp planning. They have studied various past successful as well as unsuccessful missions by other states and have planned a mission in such fashion that they could derive maximum benefits from using or modifying the existing infrastructure available with them (would have the advantage of the late beginner). Also, the cost incurred for India's Mars mission is just a fraction of their overall space budget. It is expected that India would undertake cost-effective missions also during the opportunities available in 2016 and 2018. Definitely, the planning of these missions would factor the lessons learned during the 2013/2014 mission.

It is important for India to learn from the overall experience of NASA in regard to mission planning. Over the years, NASA has completed various successful missions and there is much to learn from that. But, despite its significant scientific successes over last many years, NASA appears to be have lost its way in some fields. Stopping the space shuttle programme without having alternatives available particularity with the ISS experiment continuing is perplexing and appears to be a case of planning miscalculation. Also, due to financial compulsions, the USA had to cut the budget of NASA and this has made NASA to reorient its Mars agenda. The space shuttle example signify that starting an ambition space agenda without surety of long-term financial backing would be unproductive in long-term. For ISRO, its Mars programme is in the beginning stage. They need to plan well in advance about their future agenda and project it to the government so as to avoid any complications on a later date.

Mars is a complex issue for both scientists and technologists. Future Mars missions for various states are expected to be more complicated missions incurring significant cost. Particularly, to and fro human missions (landing on Mars) would require considerable amount of technological and financial investments. It could be imprudent for any single state to follow this agenda alone. Apart from cost, it would also be a time-consuming affair. The best option to see the human walking on Mars in near future could be to develop a global Mars programme. The space-faring and other interested states could come together and device a truly international Mars programme something akin to ISS.

Over the years, Indian space agency ISRO has gained significant reputation for its professionalism. Its space programme is viewed as one of the most successful space programmes from the developing world. Over last four decades, this organisation has succeed in increasing its stature because of the strong scientific leadership and encouraging political support by the Indian governments. ISRO has successfully developed healthy bilateral relationships with other space-faring states and also have engaged some of the non-space-faring states commercially. India actively participates on various debates at various United Nations forums in regard to space debris and space security. All this indicates that India is a proficient, respected and responsible space power.

Presently, the challenges for devising a globally acceptable space regime are far too many. Some powers at the moment are found abusing their influence at the global stage and exploiting their leadership position to shape the debate on space security. Some powers are probably engaging in a leadership role mainly to protect their self-interests. The issue of space debris is becoming more serious with increased activity and there are some predictions about the likelihood of catastrophic collisions in the space anytime in near future. At the same time, few states have already demonstrated the technologies for the weaponisation of the space making the possibility of space arms race more real. There are geopolitical challenges in respect of issues like space weapons, ASAT (anti-satellite) technologies, missile defence, etc. Hence, any proposal for negotiations to establish a space treaty/code of conduct/rules of road mechanism is going to be an extremely challenging process. There is a need to have a forum where various space-faring states could come together and at least discuss issues where no diversities exist. Such coming together could help initiation of a process of constructive engagement which in long run could assist the larger cause of space security.

It is important to remember that the geopolitical dynamics of the Cold War era where the USA and the erstwhile USSR were engaged in the space race no longer exists. The USA and Russia are working together to sustain and further develop the ISS. At the same time, they also do have differences of opinion in respect of development of space treaty architecture. The differences between these two major powers on the issue of missile defence are well known. All this indicates that modern day geopolitics allows a "space" for both the cooperation and disagreements to coexist.

Space research and technology by its very nature is an international activity. Humans are dreaming to visit and stay over Mars for long. Presently, there are significant variations in respect of the level of technological proficiency to undertake Mars missions amongst various states. Also, financial limitations are not allowing any state to undertake a major Mars programme. Hence, it could be most appropriate for the states to come together and plan a joint mission to Mars.[1] This could allow them to realise the dream of human presence over Mars in a shorter duration to time and in a lesser cost. Presently, apart for ISS, no other significant multilateral experiments are underway in space arena. The Geneva-based European Organisation of Nuclear Research (CERN) responsible for the discovery of the Higgs Boson particle in 2012 demonstrates the effectiveness of a well-coordinated multilateral scientific effort. It is important to note that in space research sector, any specific mechanism involving various states to work on a project of global interest could also emerge as one of the most effective confidence building mechanisms (CBM). In this regard, Mars offers a great opportunity.

For commencement on any multilateral effort on Mars research, there is a need for a space-faring state with wider acceptability to become proactive and initiate the process of discussions. India could take such initiative and propose the institution of an International Mars Project (IMP). ISRO could become a founding member of such project. The project should aim for the conduct of Mars human spaceflight programme within next two decades. The chief aim of the project would be not only to take the humans to Mars but also to use the process of development as an instrument to receive maximum scientific paybacks. The long duration of the project and the technological challenges would offer various socio-economic benefits and would effectively lead to deriving various scientific, technological and commercial benefits too.

India could be the most suitable candidate to lead such a project as a principal investigator and project manager. Various arguments could be given in favour of India in this regard. First, India has harmonious relations with various major space-faring states and has been collaborating with them in space arena for many decades.[2] Second, India may not be the technology leader in respect of Mars research but has efficient and globally respected scientific community which could become a pivot for larger engagement. It is also important to note that various space agencies in the world have people of Indian origin as a part of their core scientific teams. It may become easier for India to develop a global network of scientists and policy makers. Third, a project conceived by India could attract non-space-faring but important space players like Canada, Germany, South Africa and Brazil, etc., and also various global private industrial houses (aerospace industry and other related industries). Fourth, Indian state is in a position to do a reasonable financial investment for such project and also could attract finances from Indian private industries. Essentially, India could device an economic model for other states to

---

[1] In 2010, the then Russian space agency head has announced the proposal of 26 space agencies around the world joining hands to carry out a joint flight to Mars; after the year 2030, however, no additional information is available in this regard.

[2] Since the beginning of its space programme, India has collaborated with the USA and Russia (USSR) in various fields and for last few years has also been collaborating in the field of deep space missions. Space has been a flagship agenda for Indo-French relationship for many years. India has launched a satellite for Israel and have also received assistance form them for the launch of its Radar satellite. China is keen to engage India is areas like space solar power collaboration.

follow. Fifth, for both non-Asian and Asian space-faring states, India could be one of the most acceptable states to lead such project.

India should also remain prepared to face various challenges to convert this idea into a reality. There could be few major challenges for India (or that matter any other leader). First, various technological by-products could emerge from this project having strategic relevance. Because of the inherent possibilities of the materialisation of few spin-off technologies of strategic significance, some states could have reversions to join such a project. Secondly, probably technologically developed countries would not like others to become of part of the success when their astronauts reach Mars. The theme of nationalism still has significance in the era of globalisation, and hence, they would be cautious of joining such joint projects. Thirdly, it is important to sustain such projects financially for the entire duration of development and also cater for additional requirements arising in various stages of the project. There would be always concerns in regard to cost control. Within two decades, every participating state could witness roughly four governments in power (average five years term for the government in one election cycle). Financial priorities for every government could vary and impending political challenges would control its budging priorities. It would be important not only for India but also for other participating states to link this project to national prestige since beginning to keep the interest for citizens as well as governments alive for such a longer duration.

To make IMP, a reality would involve significant amount of planning and negotiation skills. Technological and financial capabilities of every participating state would dictate the nature of their role in such project. All efforts would be required to develop a mechanism so as to avoid disputes and incompatibilities. It could be important to first get the idea accepted at the political level within the country and with other anticipated participating countries. Various states and respective space agencies could be engaged towards developing a final project proposal. Also, an initiative could be taken to engage private agencies keen for both technological and financial investments. To start such a process, first ISRO would have to undertake preliminary studies on its own and develop an idea (draft project outline) for wider considerations.

The notion of dominance in space is a Cold War centric view. In the twenty-first century bracketing space only with a myopic view as a fourth dimension of warfare would be incorrect. Initiatives like IMP should not be viewed as a tool to seize any leadership role or an instrument of power projection but essentially as a means for promoting peaceful cooperation and collaboration in space. Collaborative efforts towards reaching Mars would help us not leaving the Mars agenda for future generations to concoct.

Appendices

Appendix A

Mars Missions: Time Line

A. Lele, *Mission Mars*, SpringerBriefs in Applied Sciences and Technology,
DOI: 10.1007/978-81-322-1521-9, © The Author(s) 2014

Launch Date	Name	Country	Result	Reason
1960	Korabl 4	USSR (flyby)	Failure	Did not reach earth orbit
1960	Korabl 5	USSR (flyby)	Failure	Did not reach earth orbit
1962	Korabl 11	USSR (flyby)	Failure	Earth orbit only; spacecraft broke apart
1962	Mars 1	USSR (flyby)	Failure	Radio failed
1962	Korabl 13	USSR (flyby)	Failure	Earth orbit only; spacecraft broke apart
1964	Mariner 3	US (flyby)	Failure	Shroud failed to jettison
1964	Mariner 4	US (flyby)	Success	Returned 21 images
1964	Zond 2	USSR (flyby)	Failure	Radio failed
1969	Mars 1969A	USSR	Failure	Launch vehicle failure
1969	Mars 1969B	USSR	Failure	Launch vehicle failure
1969	Mariner 6	US (flyby)	Success	Returned 75 images
1969	Mariner 7	US (flyby)	Success	Returned 126 images
1971	Mariner 8	US	Failure	Launch failure
1971	Kosmos 419	USSR	Failure	Achieved earth orbit only
1971	Mars 2 Orbiter/Lander	USSR	Failure	Orbiter arrived, but no useful data and lander destroyed
1971	Mars 3 Orbiter/Lander	USSR	Success	Orbiter obtained approximately 8 months of data and lander landed safely, but only 20 seconds of data
1971	Mariner 9	US	Success	Returned 7,329 images
1973	Mars 4	USSR	Failure	Flew past Mars
1973	Mars 5	USSR	Success	Returned 60 images; only lasted 9 days
1973	Mars 6 Orbiter/Lander	USSR	Success/Failure	Occultation experiment produced data and lander failure on descent
1973	Mars 7 Lander	USSR	Failure	Missed planet; now in solar orbit.
1975	Viking 1 Orbiter/Lander	US	Success	Located landing site for lander and first successful landing on Mars

(continued)

(continued)

Launch Date	Name	Country	Result	Reason
1975	Viking 2 Orbiter/Lander	US	Success	Returned 16,000 images and extensive atmospheric data and soil experiments
1988	Phobos 1 Orbiter	USSR	Failure	Lost en route to Mars
1988	Phobos 2 Orbiter/Lander	USSR	Failure	Lost near Phobos
1992	Mars Observer	US	Failure	Lost prior to Mars arrival
1996	Mars Global Surveyor	US	Success	More images than all Mars missions
1996	Mars 96	USSR	Failure	Launch vehicle failure
1996	Mars Pathfinder	US	Success	Technology experiment lasting 5 times longer than warranty
1998	Nozomi	Japan	Failure	No orbit insertion; fuel problems
1998	Mars Climate Orbiter	US	Failure	Lost on arrival
1999	Mars Polar Lander	US	Failure	Lost on arrival
1999	Deep Space 2 Probes (2)	US	Failure	Lost on arrival (carried on Mars Polar Lander)
2001	Mars Odyssey	US	Success	High-resolution images of Mars
2003	Mars Express Orbiter/Beagle 2 Lander	ESA	Success/Failure	Orbiter imaging Mars in detail and lander lost on arrival
2003	Mars Exploration Rover—Spirit	US	Success	Operating lifetime of more than 15 times original warranty
2003	Mars Exploration Rover—Opportunity	US	Success	Operating lifetime of more than 15 times original warranty
2005	Mars Reconnaissance Orbiter	US	Success	Returned more than 26 terabits of data (more than all other Mars missions combined)
2007	Phoenix Mars Lander	US	Success	Returned more than 25 gigabits of data
2011	Mars Science Laboratory	US	Success	Exploring Mars' habitability

Reference
http://mars.jpl.nasa.gov/programmissions/missions/log/, accessed on May 3, 2013

Appendix B

Interview with Mr. Pallava Bagla. He is a science communicator and photojournalist for more than 25 years. Presently, he is working as Science Editor for a private television channel NDTV—New Delhi Television. He is an author of the book titled "Destination Moon" which has extensively covered India's first moon mission.

1. As a science reporter you have been watching the progress of ISRO for many years. Also, you have been the author of the book on ISRO's first moon mission. This book which was published in 2008 also makes a mention about India's Mars dream. What are your views on India's forthcoming Mars programme?

Over the years ISRO has successfully developed end-to-end capabilities. Today, they can launch rockets, build satellites and undertake deep space missions. In 2008, they have successfully undertaken the Chandrayaan-1 mission. Undertaking a Mars mission is definitely the next logical step for India's space programme and needs to be welcomed. It is also important to note that the ISRO mission could cost approximately around US$100 million which is actually a meagre amount in comparison with the global standards (The NASA Curiosity Mission costs US$ 2.5 billion).

2. Since India's second moon mission has now got significantly delayed, in your opinion what would be the impact of such a delay on India's deep space agenda?

This issue needs a detailed assessment. It is important to appreciate the fact that India's first moon mission was partially successful. Against the designed life period of two years, it performed in good health only for a period of little less than one year. At the same time, this mission was successful in discovering the presence of water on the moon, which has been the major discovery of this century.

India's proposed second moon mission is a bit different from the first Mission, where an orbiter was send only to observe the moon from a distance and also this mission is a joint mission with Russia. This Chandrayaan-2 mission will have an Orbiter and Lander-Rover module. ISRO will have the prime responsibility for the Orbiter and Rover; Roscosmos, Russia will be responsible for Lander. Chandrayaan-2 is expected to be launched on India's Geosynchronous Satellite Launch Vehicle (GSLV-MkII). India is still struggling with its GSLV programme. There is a need to have a reliable GSLV available for the launch of this mission. Also, Russia is encountering some difficulties in the production of the Lander. Hence, the mission stands delayed.

It is likely that India could opt for developing the major portion of the entire mission on its own. Also, ISRO's preparations for testing with GSLV with indigenous cryogenic engine are almost complete and the vehicle is likely to be tested in 2013. Hence, there is a possibility that subsequently after doubly assuring the reality of GSLV India would undertake the launch.

Obviously, all this is delaying the second mission significantly. However, it is important to note that we have already done an Orbiter mission and the second step should be an Orbiter and Lander-Rover associated with it. It is important to demonstrate the ISRO's capability to soft-land on the lunar surface. Any future Orbiter mission would be worth the effort if some innovative payloads are designed which could help to provide new information of the moon's surface.

3. In your opinion should India collaborate with a few other States and develop a more broad-based Mars agenda instead of going solo in the future?

Post 2005 Indo-US nuclear deal, it was expected that the US would remove the four organisations of ISRO from its export control 'Entity List' immediately. However, this became possible only by 2011. Probably in 2008 during India's first moon mission, one of the reasons why various foreign space agencies including NASA participated was to work around the international sanctions regime. Now for the Mars mission since ISRO is no longer under the sanctions regime, it was not found necessary to have an involvement of the foreign element. However, it is important to note that the interaction with the various Western space agencies during the moon mission was an enriching experience for the Indian scientific community. Also, carrying various international payloads on Chandrayaan-1 has helped India to play a role towards increasing the global knowledge about the moon. A similar experiment could have been carried out in case of Mars too. But, it is important to note that international space agencies do require considerable amount of lead-time (at times even a decade) to plan, design and develop appropriate sensors. With regard to Chandrayaan-1, the idea was mooted in 1999 while actually the mission took-off only in 2008. In comparison, the Mars mission was proposed with less lead-time.

4. What is your opinion with regard to the limited payload carrying capability of this mission? Is it worth investing approximately US$ 85 million to carry such a small payload?

It is important to note that this mission is basically a technology demonstrator mission. The real challenge is to enter the orbit of Mars accurately. With a PSLV platform there are limitations on weight. Earlier, ISRO's various scientific labs had submitted nine proposals for different payloads and at that time its combined weight was expected to be around 25 kg. Subsequently, ISRO has finalised five payload proposals and they are expected to weigh around 15 kg. Hence, to conclude that ISRO is not in a position to carry 25 kg would be incorrect. In fact now they could increase the fuel by using this additional 10 kg weight. Amongst the five sensors Methane sensor is the most unique. Till date no other country has carried a Methane sensor to Mars and this particular sensor may offer some information about the possible source (origin) of Methane on Mars which could eventually lead to an answer to the question of life on Mars. Within the international community there is a great deal of excitement about India's mission and they expect this mission to offer some additional knowledge about the planet.

With regards to the criticism about the cost factor of this mission it needs to be emphasized that such arguments are very naive made with the belief that all poverty related problems in India could resolved with the US$ 100 million spend on this mission! It is important to note that India has a policy where it works towards addressing all issues regarding the social development and simultaneously also works towards research in the blue sky area which eventually leads it to playing an important role towards the development of the country. Hence, investments in space technologies are essential.

5. Do you feel that presently there is competition amongst the States in undertaking various space missions similar in nature? Like Moon and Mars missions, etc.

A broad comparison of the space programmes of India and China indicates that China is way ahead of India in almost every field in the space arena: from navigational satellites to human visits to space to undertaking space walks to moon missions. During November 2011 China's first Mars satellite, Yinghuo-1, was launched piggy-backing on the main Russian spacecraft. This Chinese mission failed because of the failure of the Russian mission. Now, there is an opportunity for India to become the first and the only Asian State (few years back Japan had attempted a Mars mission but that also was a failure) to Mars, if its mission becomes successful. Also, this could provide an opportunity for India to demonstrate its capabilities well ahead of its regional rival China.

6. As a television journalist do you feel that in India adequate amount of media converge is not given to the issues related to science and technology in general? Presently, the Indian television industry may be having more than 25 dedicated television channels for religion or spirituality but there is no TV channel for science and technology. Do you feel that in the 21st century India, TV industry in general is contributing more towards spreading harmful superstitions than making efforts for science to reach the masses? Do you feel that India's Mars agenda could be used intelligently to generate the scientific temperament in the country and cultivate rationalistic moral values?

India is a very complex society where for many the scientific awareness and faith go hand in hand. There is a possibility that the idea of dedicated TV channels for science and technology may not find many takers presently. There could be various reasons for this. It needs to be appreciated that there is need to develop a story on science and technology which could attract the attention of the masses for a longer duration of time. In the Indian context, for all these years there has been certain amount of secrecy with departments like atomic energy and space sciences. This was probably because we were under technology sanctions and hence had developed a tendency of being extra careful. For an audio-visual media like TV it is important to have access to various laboratories and places where the work is in progress. Getting correct visuals for a story is extremely important. At the same time, the media also needs to keep in mind the importance of national security issues and shoot and show only the relevant information. During the Chandrayaan-1 mission media had played an important role by showing different aspects of the mission and that had indeed caught the attention of many people. A proactive role by the media in the case of the Mangalyaan mission could definitely help ignite the minds of many people and eventually attract them more towards science.

Appendix C

Interview with Dr. Amitabha Ghosh. He currently serves as a Chair of the Science Operation Group for NASA's Mars Exploration Rover Mission. He is working in NASA for last 16 years and also was a member of the NASA Mars Pathfinder Mission. He was responsible for analysing the first-ever Martian rock which was incidentally the first rock analysed, which was from another planet.

1. Why presently different states are aiming for Mars?

Every state or a group of states (say, like the European Space Agency or ESA, which consists of the nations of Europe) has its own rationale for having a Mars programme. Essentially the people involved are keen to know more about the Red Planet.

The competing space faring bodies are aware that the engineering behind interplanetary journeys is complicated and therefore the history of reaching Mars by spacecraft has had a high failure rate (~ 66 %). To them to reach Mars is a serious challenge. They also understand that once reached it is possible through proper planning to meet with even unexpected levels of success, such as that achieved by the two Mars Exploration Rovers, which operated for years beyond their original mission specifications. Mars is also considered to be an opportunity to upgrade a nation's technological prowess

Space race was a Cold War phenomenon amongst the US and erstwhile USSR. The then US President John Kennedy was the chief architect of this idea and he had initiated the Apollo programme to demonstrate the country's technological supremacy. Modern day geopolitics doesn't position such compulsions on the space-faring States. Each space agency has its own ideas and vision about what should be their agenda in space and what should be the way forward.

2. How much do we know about the geology of Mars? What more needs to be known?

During the last 16 years we have gathered significant information about the Martian surface. Now we know much more about the entire surface and details about the nature of the landmass and other details. Due collection of such information now on the lines of Google Earth, we also have Google Mars. From the geological point of view, it could be said that the knowledge gained so far is at the level of elements and now the effort is to gain inputs at the isotope level. The mission Curiosity is expected to play an important role in this regard.

3. How would the knowledge of geology help the overall Mars agenda?

The basic aim to go to the Mars is to gain knowledge about the presence of water and life there. This information is extremely important from the point of view of establishing a human habitat in future.

4. What are the prospects of race for resources?

Since finite quantity of mineral resources are available on the earth, human beings are always in the lookout for other options and perhaps that is one of the reasons why there is talk about the mining for mineral resources on Moon and Mars. However, the ongoing technology lifecycle is unlikely to provide cost-effective solutions for bringing back any resources from other planets. Overall, technological limitations and economic unsuitability are so significant that there is hardly any possibility of any form of a race for resources. Such ideas look 'appealing' but are technologically unviable.

5. Your views about India's Mars programme

India's overall progress in the space arena is praiseworthy. Since India is for the first time attempting a mission to Mars it is obvious that it would have to view the mission more as a technology demonstrator. It appears that India wants to take the first step to check on engineering capabilities with regard to sending such a mission to Mars which is going to take 300 days to reach. This mission would provide a significant amount to experience to India's scientific community. India proposes to have a step-by-step approach and to my mind, it is the correct approach.

6. What should be India's Mars agenda for the future?

This needs to be decided by the Indian scientific community. Since they are already geared up for their first mission they must be having a definite plan for the future. The success of the first mission is likely to dictate their future actions. Once they master the technology of reaching Mars then they could concentrate on taking forward the scientific agenda.

Appendix D

Interview with Dr V. Adimurthy, Senior Advisor (Interplanetary Missions), ISRO, and Dean (R & D) at the Indian Institute of Space Science and Technology (IIST).

1. What are the major technological challenges for India's Mars mission?

The travel from Earth to Mars takes about 300 days. Mission planning, executing various manoeuvres and operations, and controlling any small deviations in its course through mid-course corrections are the challenges one will encounter for the first time. Specific challenges in reaching Mars orbit relate to power, communication and propulsion systems. Because of lower solar irradiance due the large distances, one has to provide much larger solar panel area. Onboard autonomy has to be provided as the Earth-Mars distance does not allow real-time interventions. Restart of the propulsion system, after nearly a year of travel in space, for Mars Orbit Capture manoeuvre is a major technical challenge.

One of the major challenges when the spacecraft orbits Mars is the management the S/C being in long shadows due to eclipses. The orbital plane needs to be targeted so that long eclipse durations are avoided and good illumination conditions to be ensured near periapsis. Visibility from Earth station is another matter of concern. Spacecraft attitude control for the operation of various scientific experiments is very important. All these aspects can be tackled by the mission operations team using the past experience in the spacecraft operations. Appropriate sequence of mission operations specific to Mars Orbiter Mission are planned and worked out in detail before the execution of the mission.

2. What questions on Mars Science would be addressed by India's first mission to Mars?

A number of scientific proposals are initially considered which can be classified into four categories, viz., (1) Martian atmospheric studies, (2) Solar and X-ray spectroscopy, (3) Imaging Mars, its Moons and possibly Asteroids and (4) Radio experiments. After intense discussions on the science content, maturity, readiness and mass constraints five primary payloads and a few stand-by payloads are identified. I will highlight some of the important objectives of these experiments. First and foremost is the high accuracy measurement of methane content in the Martian atmosphere and the information about its origin, whether it is biogenic or volcanic. The next important objective is the investigation of the Martian upper atmosphere escape process, especially the loss process of water. Understanding this escape process for non-magnetic planets such as Mars is very important to infer the evolutionary history of the planetary atmospheres. In another experiment the Mars Exospheric Neutral Composition Analyzer will measure in-situ the neutral composition and density of Martian upper atmosphere-exosphere at altitudes 500 km and beyond. As an additional science goal, this experiment will also provide first-ever limits to neutral particle distribution around the Mars satellite Phobos. The spacecraft will also carry a camera to capture the topographic features. Additionally, a thermal infrared multispectral imager will map the surface composition and mineralogy of Mars and will detect hot spots which indicate underground hydro thermal systems.

3. During August 2010 an ISRO team was formed under your leadership to check the viability of India undertaking a Mars programme and by November 2013 we expect the mission to be Mars-bound. What gives you confidence that in such a short span of time ISRO would be able to pull it off successfully?

We understand the complexity of the Mars Orbiter Mission we are undertaking this year. Launch scenarios and capabilities for Mars mission, detailed mission options for the future launch opportunities to Mars, Spacecraft design and configuration challenges, possible scientific experiments to augment the current understanding of Martian science, Deep Space Network Challenges are all addressed by the ISRO Team. There are several new technological challenges that I have already highlighted. All these require focussed attention. It is an intrinsic and deep understanding of the issues involved that gives us the perspectives and right directions for the appropriate solutions. During the intense deliberations of the Study Team we have discussed various kinds of Mars missions; like fly-by, orbiter and other more complex options. It appeared that a fly-by, which gives only a short time for scientific study, is not really attractive. But an orbiter or a Lander would require larger transportation capability that may not be met by our established and reliable launch systems we now have. But we then discovered that, if we can have a highly elliptic orbital mission around Mars, we can gainfully utilise our proven PSLV-XL launch system. This is an exciting opportunity, and we are delighted that very meaningful, though limited, scientific experiments can be conceived for such mission. A mission like the 2013–2014 opportunity to reach Mars is the most appropriate with the energy optimality it offers. Similarly, by a critical analysis of the other technological challenges and the solution options available to us, we come to the conclusion that we can, with a good degree of confidence, undertake this complex mission in 2013.

4. Over the years Mars has always been a tough customer for the space scientists with almost fifty percent of the missions undertaken globally being failures. What lessons has India learned from all such failures?

Of about 50 attempts so far from Earth to reach Mars, only 21 succeeded; a success rate of 42 %. A few more of them posted some partial success. Thus I agree that Mars has been a tough target for space exploration. But, I know the wind has turned in the right direction now. The modern missions have an improved success rate, with the last decade posting a success rate of about 90 %. However the challenge, complexity and the length of the missions make them very difficult; and the issues have to be properly factored in the mission and system design. Adequate testing for verifying and validating critical subsystems holds the key to success. This is particularly relevant to the propulsion system functioning for Mars Orbit capture, power management, thermal design and communication systems. All these are being carefully addressed during our preparation for the forthcoming mission.

5. For the launching of the first Mars Mission India proposes to use PSLV-XL vehicle. Could GSLV-MkIII type of a vehicle have provided a better option? Has India's limitations in the launch vehicle arena restricted it from developing a more ambitious Mars programme?

When we take up a complex mission like the Mars orbiter Mission, it is necessary to build the confidence over the strength of our proven launch systems.

Yes, the GSLV Mark III vehicle, when ready, will provide more opportunities for scientific experiments and mission options; but such a route will also come with some additional hidden costs; for example, the development of a new interplanetary orbital propulsion system which will be needed to capture the benefit of higher transport capability of the Launch Vehicle. The same is true with the regular GSLV vehicle; and these additional developmental tasks put the programme much beyond in time; and any lessons to be learnt at that stage will be very costly. I have already explained the rationale for selecting the PSLV-XL vehicle and an appropriate mission design tuned to exploit the strength of this vehicle. By designing to go into a highly elliptical orbit around Mars, with an apoapsis of 80,000 km, we open up a confident, cost-effective and faster mission option for the Mars Orbiter. Yes, in this process the scientific payload mass will be restricted; but this will give us the excellent opportunity to gain an invaluable experience in mastering the technology of this complex mission faster and cheaper and at the same time to explore a meaningful scientific objective.

6. Moon and Mars missions both belong to the category of Deep Space Missions. However, Mars offers far more challenges than a Moon mission. There is a perception that India is just upgrading its technology (almost in every field from launch vehicles to telemetry and tracking) used for moon mission for the purposes of the Mars mission. What are your views on this? Has ISRO developed any specific technologies exclusively for the Mars mission?

Obviously, in every space mission there will be several common elements, and specific new technological challenges. For example, if you take the communication system for these missions, what meets the requirements for a Moon mission will not meet those of a Mars mission; the distances in question are different by orders of magnitude. Then, at the time of Chandrayaan-1 system design itself we have exercised some amount of foresight and factored to a good extent our future requirements. Even though for the Moon mission it is technically feasible to use 14 m antenna for Deep Space Network (DSN) at the cost of some increased complexity in the spacecraft, in order to reduce the complexity onboard the spacecraft and considering the futuristic needs of interplanetary missions, 34 m antenna has been considered as the baseline. This has substantially helped in configuring DSN for our Mars orbiter mission with relatively less complex upgrades.

Another important technological challenge is the Liquid Apogee Motor operation after nearly 300 days for the capture of the spacecraft into the orbit around Mars. Chandrayaan-1 mission demanded LAM operations in two phases, i.e. the orbit-raising and lunar transfer burns near the earth in the initial 9 to 10 days and another series of burns for Moon capture and circularization after a gap of 4 to 5 days. For the Mars mission, the second phase will be after 300 days. This means that, while the basic propulsion system configuration is same, for the Mars mission we have to configure new pressure regulator protection, isolation mechanisms to improve the system safety and reliability. Such new systems have been designed by our experts and the designs have gone through the established rigorous review and testing processes.

We also must note that, while we are using the PSLV-XL launch system for the Mars Orbiter Mission, the launch trajectory to be implemented for going to Mars is entirely new and poses its own challenges in the guidance, control and ground support management. These new designs are validated by extensive simulations and testing. The very long coasting period required for this special trajectory needs a ship-borne communication terminal. It will be a temporary transportable terminal expected to be installed on an Indian Navy Ship.

Yes, I agree that the technological challenges of going to the Mars are not identical to going to the Moon. All these challenges are appropriately and sufficiently addressed in our Mars Orbiter Mission. When in future, we plan to have more scientific studies on Mars, requiring higher payloads to be taken to Mars, then I expect that our new launch systems like GSLV MkIII will be ready to meet our new planetary exploration needs.

7. During the moon mission ISRO had a mix of payloads: few Indian and few from other countries. In the case of Mars we are going solo. Any specific reasons for this? Had any of the other countries contacted India to send their payloads on the Indian craft?

Yes, the Chandrayaan-1 Mission exemplified unique international cooperation and participation in the scientific exploration of the mysteries on moon. Chandrayaan-1 payload ensemble comprises a combination of six international and five Indian scientific experiments chosen meticulously to complement each other in achieving the conceived scientific objectives. ISRO invited international space organizations to participate in the project by providing suitable scientific payloads.

In view of the stringent payload constraints, it is not possible in our Mars Orbiter Mission to make the Announcement of Participation to other interested countries. Since we have not extended such invitation, there were no proposals submitted either. From a large list of about 20 proposals from our Indian scientific community, and after a rigorous review process five experiments are selected for this mission.

8. In terms of payload, this mission would carry only five scientific instruments and the total weight of this entire payload would be 13.5 kg which is slightly more than the weight of a nano satellite. Initially, there was a plan for more number of payloads (7 to 8 payloads with total weight of 25 kg) but finally ISRO has settled for five. Do you feel that the entire effort is worth only such a small payload?

The total mass of the spacecraft that will be orbiting the Mars is about 500 kg. The spacecraft bus system weighs around 486 kg, comprising of many subsystems that ensure the reliable and safe operations of the spacecraft like; propulsion, structure, thermal, control, inertial, battery, power electronics, solar panels, data handling and other such systems. Considering all the engineering uncertainties in all these necessary systems, we have an assured mass for scientific payloads of around 14 kg. Thanks to the maturity and the technology of miniaturisation of the scientific payloads, we can accommodate at least five carefully selected payloads. The composition and coverage of scientific investigations involved in these instruments is very substantial. So, from the point of technological challenge and innovations as well as the scientific content, the Mars Orbiter Mission will make a significant statement and impact.

9. How do you rate India's mission in comparison with other missions undertaken so far? Are we reinventing the wheel? What new knowledge would India's mission offer to the rest of the world?

Since we are optimally utilising existing resources, technologies and have made an innovative mission design in tune with the resources, our Mars orbiter mission will be among the fastest and cheapest missions ever undertaken. As far as the expected science output is concerned, we have carefully chosen our experiments at Mars to address the gaps in the existing knowledge; so the outcome of these efforts will be very interesting to the world science community, and may even bring out a path breaking discovery.

10. What are the challenges of Mars orbital insertion? Based on the Japanese experience with the Nozomi mission, does ISRO have a plan 'B' ready for its 2013–2014 Mars mission?

JAXA's Nozomi mission of 1998 had primarily failed because of a malfunction of a valve in the propulsion system which resulted in loss of fuel. This is an important lesson, and as I have already emphasized earlier, ISRO is ensuring valve isolation and redundancy through which the long operations of LAM are made reliable and safe by design. We are also testing the system extensively to ensure the propulsion system performance.

Nozomi mission had planned several Earth swing-bys and Lunar swing-bys to get extra energy benefit at the cost of mission concept complexity. During the Mars Mission studies conducted by us, we are indeed aware of the benefits to be accrued by attempting the Lunar and Earth swing-bys in the mission profile; but we have taken the conscious approach not to make the mission very complex in the very first attempt. It may also be noted that Nozomi's on-orbit dry mass was around 260 kg with a propellant of 290 kg. Here our Mars Orbiter Mission is in a relatively comfortable position with a dry mass of about 500 kg with a propellant loading of 850 kg.

After the initial valve malfunction that left the Nozomi spacecraft short of fuel, JAXA worked out a new plan for Nozomi to remain in heliocentric orbit for four more years, and after two Earth fly-bys to reach Mars; but this also could not be implemented successfully because of additional problems the spacecraft encountered. So, the mission ended with just Mars flyby in December 2003. Regarding a Plan-B for our Mars Orbiter Mission, if such a contingency unfortunately arises due any problem, the recovery plan, as done in Nozomi, will be a function of the state of our spacecraft after encountering a problem; so such contingency plan can only be worked at that point of time analysing all the possible options that may be available.

11. After the launch of the mission it would take 300 days to reach Mars. How would ISRO monitor and control the mission for all these days? Would it be possible to undertake any midcourse correction, if required?

Yes, certainly the Mars Transfer Trajectory (MTT) will be continuously monitored, and mid-course corrections will be made based on the orbit determination of the MTT. We have apportioned appropriate fuel in the fuel budget to make the mid-course correction to compensate for the expected normal deviations in placing the spacecraft in the MTT. It may be recalled that such a provision was also made in our Chandrayaan-1 mission, and the actual mid-course correction imparted was very much smaller than the provision made, as the insertion to the Lunar Transfer Trajectory was very accurate.

12. Keeping real-time contact with the Mars mission is not possible. What plans are in place to cater for the significant amount of time-lag of 40 min?

During the Mars Orbiter Mission, the spacecraft can reach a farthest distance of 400 million kilometres. At this distance, it will take as long as 20 min for a signal to travel from the ground station to the orbiter and another 20 min from the spacecraft to the ground station. This means that spacecraft operations cannot be made by on-line real-time command and monitoring from the ground station. This has to be handled by providing required autonomy in the spacecraft operations and where necessary by providing in design for delayed execution of spacecraft operations based on the analysis from the Mission Operations Complex on the ground and up-linking the strategy to the spacecraft.

13. Which countries are involved in assisting India in respect of deep space network? Is Indian Navy also involved?

International co-operation and collaboration are certainly needed for the ground station support during the launch segment; and also during the orbit raising, transfer trajectory to Mars and orbital operation phases. I have already mentioned the need for ship-borne terminal during the long coast of the launching phase. For specific MOUs entered with other countries in this respect, I request you to get the authentic inputs from the Scientific Secretary, ISRO.

Appendix E

Interview with Dr K. Radhakrishnan, Chairman of the Indian Space Research Organisation (ISRO). He started his career as an avionics engineer in 1971 at ISRO's Vikram Sarabhai Space Centre and took charge as Chairman ISRO on 31 October 2009. He has obtained his PhD from India Institute of Technology (IIT)

Kharagpur and has also earned his post graduate diploma in management (PGDM) from the Indian Institute of Management (IIM) Bangalore.

1. What is the genesis of India's Mars programme?

Mars holds a very special position in the quest for planetary exploration, as the conditions in Mars are believed to be hospitable and the planet being similar to Earth in many ways. The surface of Mars has many geological features that have recognizable counterparts on Earth. Evidences suggest that, Mars once had rivers, streams, lakes, and even an ocean. Recent discovery of Methane on Mars suggest that life could exist on Mars. But, the question yet to be answered is whether Mars has a biosphere or ever had an environment in which life could have evolved and sustained. As in any field of scientific exploration, we need to learn more and more about the Martian surface, its topography, geology, landforms, mineralogy, its upper atmosphere etc.

Advisory Committee on Space Sciences (ADCOS) identified mission to Mars as one of the important components of planetary exploration for India. ISRO has been conducting detailed technical and scientific studies for undertaking an Orbiter mission to Mars, since August 2010. Study Report indicated that it is possible to undertake Mars Orbiter mission using Indian Launch Vehicle. Also, having demonstrated the technological capability in reaching the Moon, the next logical destination in planetary exploration is Mars.

The Indian Mission to Mars would establish Indian technological capability to reach the Mars, orbit around it and also provide an excellent opportunity, to the scientific community, to further understand the Martian Science.

2. **Is the name Mangalyaan an official name for the November 2013 mission?**

As of now, the mission is called as Mars Orbiter Mission. It is not yet named Mangalyaan or otherwise.

3. **India's Mars mission has been viewed as a technology demonstration project and it appears that the 2013 mission has a very limited scientific agenda. Should ISRO have waited for a few more years and opted for a 2016 launch?**

Being the first Indian mission to the planet Mars, the primary technological objective is to realize a spacecraft with a capability to reach Mars (*Martian Transfer Trajectory*) and then to orbit around Mars (*Mars Orbit Insertion*). There are many technological challenges, which need to be negotiated like *providing augmented radiation shielding to the spacecraft for its prolonged exposure; building high level of onboard autonomy within the Orbiter to deal with communication delay of the order of 40 min; robustness and reliability of propulsion system, which needs to work again after almost 300 days voyage; Martian orbit insertion* etc. Our primary concern is to send a spacecraft from Earth to Mars with the least amount of fuel possible. However, it is also planned to undertake few scientific studies during the orbital life of the spacecraft, using Indian instruments onboard to study *the Mars surface features, constituents of*

Martian atmosphere like methane, dynamics of upper atmosphere of Mars etc. The mission would also provide multiple opportunities to observe Mars moon, "Phobos".

Coming to the point of launch opportunities, we need to understand that, to travel from one planet to another, the two bodies must have appropriate angular geometry for minimum energy transfer. Such opportunity recurs periodically, which for Earth–Mars is about 780 days (approx. 26 months). We have studied the detailed trajectory design for placing a spacecraft (*with limited payload capacity*) in an orbit around Mars using PSLV-XL and GSLV.

Considering the minimum energy transfer and providing a meaningful mission life, the earliest two opportunities using PSLV-XL are in the year 2013 and 2018. After November 2013, the next launch opportunity to Mars exists only in January 2016. However, the January 2016 launch opportunity is not as energy efficient as the November 2013, with a result, launch of an Orbiter around Mars isn't feasible using PSLV-XL in 2016.

Moreover, ISRO has demonstrated in Chandrayaan-1 mission that, an unmanned mission to a celestial body could be realized without major modifications to the existing configuration of Polar Satellite Launch Vehicle (PSLV-XL), which is a proven launcher. The Mission to Mars will place India amongst select few countries which have achieved this rare feat.

4. **Would India have an ongoing Mars programme or would the 2013 mission be a one-mission agenda? If not, then has ISRO developed any well articulated Mars agenda? Has ISRO started preparations for any missions for the launch windows in 2016 and 2018 respectively?**

As I explained earlier, reaching to Mars and capturing Martian orbit is a very complex mission involving many technological challenges. We must also understand that the success rate of Mars mission has been about 50 % globally. Since 1960, nearly 42 un-manned missions to Mars have been undertaken. Of these, nearly 21 missions have failed and the failures have been mostly due to the innate complexities and duration of missions to Mars.

Thus, presently, we are focussing on the realisation of a successful mission to Mars. Of course, while exploring the launch opportunities in the year 2013, 2016 and 2018, we have also studied the possible launch opportunities, corresponding velocity requirements and payload capabilities for Orbiter missions to planet Mars beyond 2018 as well.

5. **For India's second moon mission the lander would be supplied by Russia. How close is the Indian scientific community from making indigenous rovers and landers and other robotic instruments which could be used in future deep space missions?**

At present, we are studying the feasibility of developing Lander in India. The development of Lander as well as the landing on the lunar surface involves many complex technologies. To identify the safe location for landing, we need precise estimation of lunar gravity; lander should take real time pictures, process it on board and command the lander module for safe landing. We are confident that we could do it as many of these mechanisms like *propulsion systems, remote sensing, guidance systems* etc. are part of our space activities. As far as development of Rover is concerned, we have made significant progress. The design and development of the 4 wheel Rover Bread Board Model has been completed and mobility activities are successfully demonstrated. Presently, modification of the 4 wheeled Rover to 6 wheeled Rover is in progress.

6. Are there any discussions/proposals/plans to undertake missions to other planets or asteroids?

The observation of astronomical objects is definitely a part of our Space Science Programme. Towards this, we plan to launch a dedicated Indian Astronomy Satellite, ASTROSAT in 2014, which would enable multi-wavelength observations of the celestial bodies in X-ray and UV spectral bands simultaneously. The world astronomical community is looking at this satellite as in the next five years there are not many astronomy satellites planned by rest of the world. As part of planetary exploration, at present, we are planning study of Moon and Mars.

Technically speaking, ISRO has the capability to launch satellites to reach nearby planets. However, the large distances, harsh environments for the instruments to function and produce data are challenging tasks. For missions to outer planets, power generation is the main concern since it gets further away from the Sun and requires RTG capabilities. All these are at present being discussed as challenging developmental areas, rather than missions.

7. Do PSUs and private organisations have any role in India's Mars agenda?

ISRO, right from the beginning, has been encouraging the participation of Indian Industry, both public and private, towards achieving shorter turnaround times for realising its missions. Around 500 Indian industries are participating in realising variety of hardware, components, sub systems and systems, stage propellant tanks, fabrication activities etc. More than 160 types of electronic components and 115 types of space grade materials have been developed and qualified for use in spacecraft, launch vehicles and ground systems. Industry consortium partners have continued supplying VIKAS engines for PSLV and GSLV stages, meeting the propellant requirements of the space programme. Industry participation is on the rise in area of spacecraft power system, spacecraft building and related activities.

Coming specifically to the Mars Orbiter Mission, the primary mechanical structure and the equipment panels are fabricated and delivered by Hindustan Aeronautics Ltd as per ISRO's design. High Gain Antenna Reflector mold fabrication, power electronics systems fabrication and few of the spacecraft communication systems have also been realised through industries. Indian Deep Space Network (IDSN) station located in Byalalu, Bangalore was established with a view to meet, not only the requirements of Chandrayaan-1 mission, but also

ISRO's future deep space missions. IDSN is being augmented to support Mars mission with industry participation.

8. What are the strategic advantages of deep space missions?

Deep space missions have inspired generations of scientists and technologists all over the world. It has contributed to immeasurable technological breakthroughs in many areas and yielded intangible benefits to the mankind. Critical to the deep space exploration are the development of advance technologies in reliable spacecraft systems, propulsion systems, robotics systems, communication and guidance systems etc.

The deep space missions are exciting, challenging and drives technology development in terms of instruments and capability to reach there. Some ingenuity came out of the Chandrayaan-1, it made us build new instruments, propulsion systems and the Deep Space Network. The Mission to Mars will also upgrade our technological capability and place India amongst select few countries which have achieved this rare feat. It would pave the way for future scientific missions, and bring a strategic advantage to India in the international decision making process on matters related to Mars and deep space. It would also generate national pride and excitement in the young minds.

9. The Human mission to Mars could be viewed as a global dream. Do you feel that various countries should join hands to achieve this dream collectively? Particularly, far too many financial and technological challenges would be taken care of, if such an idea becomes a reality. Should ISRO propagate such an idea and take a lead to bring the international players together?

Applications of Space technology for the benefit of mankind are our thrust unlike, say, Russia, the United States or China which are after human space flights, space stations and such activities. As on today, Human mission is not in our space agenda. We are in very early phase of developing few critical technologies required for realising a human mission. Certainly, human mission to Mars will be very costly affair, difficult for any single country to afford and realise all the technologies on its own. This may turn global competition into global collaboration and in future, synergy may emerge among space faring nations to achieve such missions collectively.

About the Author

Wing Commander Ajey Lele is a Research Fellow at Institute for Defence Studies and Analyses, an Indian think-tank on security and strategic studies. He is a postgraduate in physics and has obtained his doctorate in international relations. He works on issues related to Weapons of Mass Destruction (WMD), Strategic Technologies and Space Security. He is an author of book titled Asian Space Race: Rhetoric or Reality? Published by Springer.

A. Lele, *Mission Mars*, SpringerBriefs in Applied Sciences and Technology, DOI: 10.1007/978-81-322-1521-9, © The Author(s) 2014

About the Author